Applying Ecosystem and Landscape Models in Natural Resource Management

Robert E. Keane

USDA Forest Service
Rocky Mountain Research Station
Missoula Fire Sciences Laboratory
Missoula, Montana

CRC Press
Taylor & Francis Group
Boca Raton London New York

CRC Press is an imprint of the
Taylor & Francis Group, an **informa** business

A SCIENCE PUBLISHERS BOOK

CRC Press
Taylor & Francis Group
6000 Broken Sound Parkway NW, Suite 300
Boca Raton, FL 33487-2742

First issued in paperback 2021

© 2019 by Taylor & Francis Group, LLC
CRC Press is an imprint of Taylor & Francis Group, an Informa business

No claim to original U.S. Government works

Version Date: 20190709

ISBN 13: 978-0-367-77929-0 (pbk)
ISBN 13: 978-0-367-34000-1 (hbk)

Visit the Taylor & Francis Web site at
http://www.taylorandfrancis.com

and the CRC Press Web site at
http://www.crcpress.com

*I would like to dedicate this book to my wife
Liz – my favorite traveling partner*

Acknowledgements

I would like to thank the following people for providing thorough reviews of each of my chapters: Geoff Cary, Australian National University; Kristen Emmett, Montana State University; Mike Flannigan, University of Alberta; Eric Gustafson, US Forest Service; Lisa Holsinger , US Forest Service Rocky Mountain Research Station; Signe Leirfallom, Blackfoot Challenge; Rachel Loehman, USDOI US Geographical Service; Carol Miller, US Forest Service Rocky Mountain Research Station; Kellen Nelson, Desert Research Institute; Sean Parks, US Forest Service Rocky Mountain Research Station; Eric Henderson, US Forest Service Northern Region

Preface

Managing today's lands is becoming an increasingly difficult task. Complex ecological interactions across multiple spatiotemporal scales create diverse landscape responses to management actions that are often novel, counter-intuitive and unexpected. To make matters worse, exotic invasions, human land use, and global climate change further complicate this complexity and make past observational ecological studies limited in application to the future. Natural resource professionals can no longer rely on empirical data to analyze alternative actions in a world that is rapidly changing with few historical analogs. New tools are needed to synthesize the high complexity in ecosystem dynamics into useful applications for land management.

Some of the best new tools available for this task are ecological and landscape simulation models. Few empirical databases and statistical analyses can account for the great environmental variability and complexity across the appropriate time and space scales needed for management applications. Natural resource management will need to integrate ecosystem simulation modeling with conventional analysis to make better decisions in our uncertain future. Land management professionals need to realize that ecological modeling must become an integral part of their assessment activities and embrace modeling as they have embraced the cell phone and personal computer.

This realization will probably come at a great cost. Many land management professionals and scientists have little expertise in simulation modeling, and the costs of training these people will probably be exorbitantly high because most ecosystem and landscape models are exceptionally complicated and difficult to understand and use for local applications. Highly technical expertise in remote sensing, spatial analysis, statistical modeling, data management, and GIS is often needed to prepare and run ecosystem models and to understand their output, especially those models that are implemented in a spatial domain. These talents are rare across many resource professionals because modeling has never been a major part of their job. However, there are a few key concepts that most novice modelers can quickly learn to employ complex landscape models for local land management analysis in the future.

This book was written to provide natural resource professionals with the rudimentary knowledge needed to properly use ecological models and then to interpret their results in the appropriate context. It is based on the lessons learned from a career spent modeling ecological systems. The book is not intended to be a review of ecosystem models nor a modeling primer as there are plenty of these books already written. Rather, it is intended as a reference for novice modelers to learn how to correctly employ ecosystem landscape models in natural resource management applications and to understand subsequent modeling results. This book mostly deals with landscape and ecosystem simulation models, which are models that simulate ecological processes over time in a spatial domain. Stand-based models are discussed but not emphasized because most have already been integrated into landscape models. However, many of the principles and recommendations in this book could be applied across many other classes of ecosystem models from point, stand, landscape, regional, and global models.

The first chapter identifies the need for modeling and the objectives of the book. It is in the second chapter where the generalized modeling terminology is introduced as context for the rest of the book. Various types of ecological models are discussed to provide the context for the main portion of the book. Then the phases of a model project are detailed in separate chapters. Chapter 3 describes the creation of a modeling project from the statement of modeling objectives to selecting the model and identifying various spatial data requirements. Chapter 4 deals with the initialization of the model, or what is needed to describe the starting conditions for the model. Next is a chapter on model parameterization or quantifying various coefficients used in model algorithms (Chapter 5).

Chapter 6 concerns the laborious process of calibrating the model to ensure some degree of realism, while Chapter 7 is dedicated to the validation or evaluation of model results. Chapter 8 describes how to execute the model for implementation of the modeling project. Then there is a chapter that discusses the analysis of modeling results in both the calibration phase and the final phase of a project. The book ends with a chapter on the general barriers, limitations, and problems that may occur over the course of the modeling project.

The main goal of this book is to provide a generalized guide and reference for using landscape simulation models in natural resource management applications specifically for the manager, researcher, and resource professional with little to no experience in modeling. It is my sincere hope that the confidence level of novice ecosystem modelers will be bolstered by this book, and they will embrace simulation modeling as another powerful tool in the land management garage. I also hope that the material in this book will help create more accurate, realistic, and understandable model results for specific modeling projects. I firmly believe that the only way to become a proficient ecosystem modeler is to do it all the time, and most people, especially those in natural resource management, will have many other things to do other than modeling. Therefore, these people can rest assured that this book will be available to remind them of various tasks needed to successfully complete a modeling project.

Contents

Acknowledgements   v

Preface   vii

1. Introduction—Who Needs This Book?   1
2. Modeling Fundamentals—What You Need to Know to Use This Book   11
3. Project Design—How to Plan a Modeling Project   53
4. Initialization—How to Begin a Simulation   80
5. Parameterization—How to Tune the Model for Local Applications   94
6. Calibration—How to Tuning the Model for Realism   113
7. Validation—Determining Model Uncertainty   133
8. Execution—Implementing the Model Project   157
9. Analysis—Evaluating Model Results   165
10. Issues—Things to Think About When Using Models   188

Index   207

1

Introduction

Who Needs This Book?

"The computer is not, in our opinion, a good model of the mind, but it is as the trumpet is to the orchestra—you really need it. And so, we have massive simulations in computers because the problem is, of course, very complex."

Gerald Edelman

An ecosystem model—an abstract, usually mathematical, representation of an ecological system (ranging in scale from an individual population, to an ecological community, or even an entire biome), which is studied to better understand the real system (Hall and Day 1977).

ABSTRACT

Predicting what will happen as a consequence of management actions, or lack of action, is a fundamental goal of landscape planning. Historically, land management professionals used results from empirical studies, coupled with their own expertise and wisdom, to make these predictions. But then climate change made most of the findings of field studies and the accrued wisdom of professionals over the last 100 years somewhat limited for the management of tomorrow's landscapes. To fill this void will be ecosystem simulation models, which may become critically important tools for managing landscapes in the future because they synthesize highly complicated ecological interactions into complex computer programs that can be used to simulate alternative management actions. This book was written for those people who are unsure of how to properly develop and implement a modeling project to assess the impacts of management actions in local to regional applications. It is meant to be used as a generalized manual for employing ecological models to solve resource management problems. This chapter describes why this book is important and who needs to use this book.

Why Do We Need Modeling?

So much of natural resource management depends on scientifically credible projections of future conditions under both passive and active management. Predicting what will happen as a consequence of management actions, or lack of action, is a fundamental requirement of the planning process. Historically, land management professionals used results from empirical studies, coupled with their own expertise and wisdom, to make these predictions. But then something happened. The rapid influx of carbon dioxide and other greenhouse gases into the atmosphere from human activities over the last 100 years has changed everything (IPCC 2007). Changes in the earth's climate systems because of greenhouse warming are now rapidly creating new climate futures that have no analog in the recent past (Flannigan et al. 2009, Fei et al. 2017), and as such, most of the findings of empirical studies and the accrued wisdom of professionals over the last 100 years may not be entirely valid for the management of tomorrow's landscapes (Gustafson 2013). Yet, the valuable information gained from results of past studies should never be tossed in the trash bin or relegated to distant archives as these results still have great value to management. They do, however, need to be interpreted in a brand new context and used in other ways. No longer can we assume that statistical correlations and empirical analyses done in the past will hold in our uncertain climatic future to effectively manage tomorrow's resources (Scheller 2018). We need to integrate past study findings into a tool that is based on physical, chemical, and biological foundations to predict the consequences of alternative land management actions under various climate futures. And that tool has been around for the last 50 years—ecosystem simulation modeling: the computation of ecological responses over time using a computer program.

Ecosystem simulation models will be a critically important tool for managing landscapes in the future for a number of reasons. First, models are needed because ecosystems are remarkably complicated. Myriad interacting ecological processes result in complex biotic and abiotic responses (Bachelet et al. 2000, Allen 2007, Bockino 2008). It is nearly impossible for one person to understand the complexity of all possible ecological interactions, nor is it possible to collect enough data on these interactions to completely understand their consequences. Models also provide a means to synthesize state-of-the-art knowledge, current research findings, and general information into a tool that explicitly recognizes interactions and predicts various ecosystem responses as a result of changing conditions. Next, models can be used to extrapolate spotty empirical data over larger areas and for longer time spans to provide greater spatiotemporal scope for management decisions. Models also identify those ecological processes that are poorly understood and need further research. Comparisons of alternative management actions are generally thought to be the greatest strength of ecological modeling in natural resource management, but models can also be used for many other phases of management, from planning to real-time decision making, risk and hazard analysis, prescription development, and treatment prioritization, and they can be used across the many scales of management actions (Figure 1.1). Most importantly, models can be used to evaluate ecosystem and landscape responses as novel climates and disturbance regimes change over time. Many are now finding that describing landscape and ecosystem response as climates change is best done with ecological simulations (Loehle and LeBlanc 1996, Bachelet et al. 2003). While models are far from perfect, they may be the best tools we have for future land management.

Temporal Scale		Spatial Scale					
		Stand	Landscape	Region	Stand	Landscape	Region
		Use of models over spatial scales			Use of models over temporal scales		
	Days to Years	*Evaluate Risk*			*Prescribe activities*	*Locate treatments Units*	*Prioritize areas and allocate resources*
	Decades	*Schedule Treatments*					
	Centuries	*Provide Targets*					
		Types of models			Examples of models		
	Days to Years	*Growth and yield, community, stand, gap models*	*Landscape or watershed models*	*DGVM – Dynamic global vegetation models*	IBM	ABM	Climate models
	Decades				FVS-FFE	TOPMODEL	BIOME-BGC
	Centuries				JABOWA	LANDIS, FireBGCv2	LPJ-spitfire

Figure 1.1. A summary of the ways ecological models can be used in natural resource decision-making across temporal and spatial scales. Management decisions are often made at three broad spatial scales—stand (< 100 ha), landscape (< 200,000 ha), and region (> 200,000 ha)—and at three temporal scales—years, decades, and centuries. There are specific uses of models for each combination of space and time scale. Adapted from Reinhardt et al. (2001). Citations for models: ABM-agent based models as reviewed by An (2012); Biome-BGC-(Running and Hunt Jr. 1993); Climate Models-operational climate forecast models; FireBGCv2-Keane et al. (2011); FVS-FFE-Beukema et al. (1997); IBM-Individual based models such as Butler (2003); JABOWA-Botkin and Schenk (1996); LANDIS-Mladenoff (2004); LPJ-Spitfire-Bachelet et al. (2003); TOPMODEL-Beven and Freer (2001).

Why is This Book Needed?

This book was written for those people who are unsure of how to properly develop and implement a modeling project to assess the impacts of management actions in local to regional applications. It is meant to be used as a generalized manual for employing ecological models to solve resource management problems. This book is not an introduction to modeling—the literature is teeming with books on that subject. It is also not a manual for a specific model nor does it describe the use of models for different ecosystems, and it is not a handbook for building models—that would be a difficult and complicated task and far beyond the scope of only one book. More importantly, this book is not a comprehensive literature review of existing models and it is not a comparison of ecosystem models, as both would require a specific context in which to base the evaluation or comparison. This book was written with the primary purpose of providing general guidance on executing a modeling project for common natural resource management problems. I've found that, in many of the modeling projects, most people are not completely confident in how they prepared the model's input and how they interpreted the output. This book is for them. If you are someone who has been assigned to evaluate landscape responses to management activities and you have decided that modeling is the appropriate tool but you have limited expertise in modeling, then this book is for you. In summary, the goal of this book is to provide enough information on how to employ a model for a natural resource management project so that the user has high confidence in the modeling results and is able to interpret model results in the proper context.

What Does This Book Contain?

It is important that the reader know that this book only deals with ecological simulation models, and specifically emphasizes those models that are implemented for terrestrial ecosystems in a spatial domain. It does NOT cover statistical modeling using multivariate analysis, although many ecological models have statistical sub-models embedded within their structures. Examples of statistical models are timber growth and yield models for simulating stand timber volume changes, species distribution modeling to determine species current and future ranges, and phenomenological modeling relating empirical data to measured biophysical gradients often using multivariate modeling. Moreover, this book does not cover coarse scale ecosystem models, such as Dynamic Global Vegetation Models (DGVMs), as these models simulate dynamics over spatial scales that are rarely used by resource managers.

There are many computer models that simulate various environmental concerns for natural resource management. These include hydrological models, wildlife models, tree growth models, aquatic ecosystem models, and a host of other resource models. This book deals mostly with a special

5

class of ecological models that are often applied when solving natural resource management issues—**Landscape Ecosystem Simulation Models (LESMs).** The word "landscape" in LESM is used to denote that these models simulate processes in a spatial domain, which may or may not include spatially explicit models that directly simulate spatial processes such as seed dispersal and fire spread. The word "ecosystem" is used to emphasize that these models simulate multiple interacting ecological processes, and the word "simulation" is used to signify that the ecological processes are modeled over time. And last, the word "model" is used to represent that these are computer programs written to synthesize complex ecological ecosystem behaviors into algorithms that are programmed in a specific computing language. While LESMs are the central theme of this book, nearly all of the material would also apply to other non-spatial ecosystem simulation models (e.g., stand models) and other landscape simulation models that ignore spatial processes.

In general, the models that this book covers simulate four basic factors for terrestrial systems—climate, vegetation, disturbance, and management actions—at various resolutions and detail. *Climate* is important because it is the fundamental top-down process to which all ecosystems respond. However, due to the sheer number and incredible diversity of climate models, this book will only cover the types of data used in ecological models to represent climate; this book does not cover how to use climate models or how to select climate data sets. *Vegetation* is the major biota that is primarily responding to top-down processes of climate and it is also providing the energy and habitat for all other biota. *Disturbances* are perturbations that depend on climate and vegetation, and in turn, modify the two. Last, *management actions* are the primary feedbacks between ecosystem dynamics and human land use, and they provide managers the ability to craft alternative scenarios in natural resource management—the primary context of this book.

As mentioned, this book is for the person or team of people that is at the beginning of implementing a modeling project for natural resource management. A generalized set of steps to complete any modeling project is found in Figure 1.2. The first step is to become familiar with modeling terminology and fundamentals to better understand the modeling process (Chapter 2). Then, the preliminary modeling project design steps, such

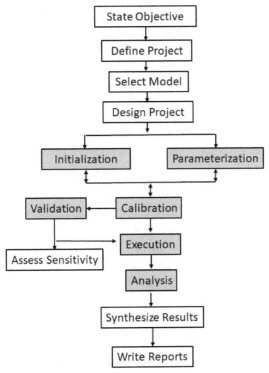

Figure 1.2. A generalize flow chart of all the steps involved in the planning, design, and implementation of a ecological modeling project as covered by this book. At the center of this chart are six critical phases in a modeling project (light gray) that form the critical chapters in this book. The remaining steps (white) are detailed in the context of the six critical phases.

as setting an objective, forming the sideboards, selecting a model, defining the landscape, and designing the project are needed to develop a blueprint for the project (Chapter 3). Once the modeling project has been designed, the six essential phases of modeling must be completed (Figure 1.2). *Initialization* concerns quantifying the values of the starting conditions in the selected model (Chapter 4) and *parameterization* is quantifying the values of all parameters in the model (Chapter 5). Once these are done, it is necessary to *calibrate* the model to produce realistic output by adjusting input parameter and initial conditions (Chapter 6). Next, all modeling projects need some sort of evaluation to determine the quality and accuracy of simulated results, so Chapter 7 details steps used in how to *validate* model behavior using extensive evaluations. Then, the model must be *executed* to implement the project's design (Chapter 8). Once all simulations are completed, the output must be *analyzed* to answer the project's objectives. And last, there are various issues that are often encountered during any modeling project and information on how to address these issues is presented in Chapter 10. Readers of this book should be able to implement a comprehensive and successful modeling project using material from this book.

This book may appear to be a linear, step-by-step guide for implementing ecological models, but many phases presented in this book could overlap in some modeling projects. Therefore, it is suggested that the reader scan the entire contents of this book before starting a modeling project. There are suggestions and material in some chapters that refer back to previous chapters; analysis software, for example, is used in the validation, calibration, and analysis phases, but only discussed in Chapter 9.

What Doesn't This Book Contain?

There are several caveats concerning the material in this book. While there is ample discussion on how to implement a successful modeling project, the reader may find that some chapters don't include enough detail to help them actually complete the steps presented. This is primarily because it is extremely difficult to present a specific procedure or protocol for all modeling projects because each modeling project is inherently unique due to local conditions, specific modeling objectives, and the model used. It is difficult to craft a comprehensive set of steps for each phase without knowing the details of the project, such as what data are to be generated, what actions will be simulated, and what area is being simulated. This book should provide the initial steps for starting, implementing, and reporting a modeling project, but it does not give detailed protocols for each phase of the modeling process. Readers must figure out many of these steps on their own.

This book was written specifically for professionals in the natural resource management field, and as such, often does not contain detailed modeling procedures that many ecological modelers often find in the literature. I purposely left out many citations in the interest of being brief rather than include a broad literature review specifically for the modeling audience. Hopefully, the level of detail in this book is sufficient for implementing a natural resource management modeling project, but it does not contain exhaustive reviews of the literature for every step or phase presented. While this book was written for the ecological professional, there should be enough information for all people in the natural resource management and research to guide the construction of a modeling project.

Admittedly, there are several aspects in this book that are heavily biased to my area of expertise—the simulation of vegetation, climate, and disturbance dynamics across terrestrial landscapes. Many examples and references in this book are taken from that diverse field. Moreover, most of my modeling research has been done in the Pacific Northwest and northern Rocky Mountains of the United States. As a result, this book often uses examples that are taken from western North American conifer forest ecosystems and most disturbance examples concern wildland fire. This is a direct result of my familiarity with those systems and that disturbance. I hope that these examples will be understandable by those outside of this narrow scope.

References

Allen, C. D. 2007. Interactions across spatial scales among forest dieback, fire, and erosion in Northern New Mexico landscapes. Ecosystems 10: 797–808.

An, L. 2012. Modeling human decisions in coupled human and natural systems: Review of agent-based models. Ecological Modelling 229: 25–36.

Bachelet, D., James M. Lenihan, Christopher Daly and R. P. Neilson. 2000. Interactions between fire, grazing and climate change at Wind Cave National Park, SD. Ecological Modelling 134: 229–244.

Bachelet, D., R. P. Neilson, T. Hickler, R. J. Drapek, J. M. Lenihan, M. T. Sykes, B. Smith, S. Sitch and K. Thonicke. 2003. Simulating past and future dynamics of natural ecosystems in the United States. Global Biogeochemical Cycles 17: 1045.

Beukema, S. J., Jason A. Greenough, D. C. E. Robinson, Werner A. Kurtz, Elizabeth D. Reinhardt, Nicholas L. Crookston, James K. Brown, Colin C. Hardy and A. R. Stage. 1997. An introduction to the Fire and Fuels Extension to FVS. pp. 191–195. *In*: Proceedings: Forest Vegetation Simulator Conference. United States Department of Agriculture, Forest Service, Intermountain Forest and Range Experiment Station, Ft. Collins, CO USA.

Beven, K. and J. Freer. 2001. A dynamic TOPMODEL. Hydrological Processes 15: 1993–2011.

Bockino, N. K. 2008. Interactions of white pine blister rust, host species, and mountain pine beetle in whitebark pine ecosystems in the Greater Yellowstone. M.S. Thesis. University of Wyoming, Laramie, WY.

Botkin, D. B. and H. J. Schenk. 1996. Review and analysis of JABOWA and related forest models and their use in climate change studies. NCASI Technical Bulletin Number 717.

Butler, M. J. 2003. Incorporating ecological process and environmental change into spiny lobster population models using a spatially-explicit, individual-based approach. Fisheries Research 65: 63–79.

Fei, S., J. M. Desprez, K. M. Potter, I. Jo, J. A. Knott and C. M. Oswalt. 2017. Divergence of species responses to climate change. Science Advances 3.

Flannigan, M. D., M. A. Krawchuk, W. J. de Groot, B. M. Wotton and L. M. Gowman. 2009. Implications of changing climate for global wildland fire. International Journal of Wildland Fire 18: 483–507.

Gustafson, E. 2013. When relationships estimated in the past cannot be used to predict the future: using mechanistic models to predict landscape ecological dynamics in a changing world. Landscape Ecology 28: 1429–1437.

Hall, C. A. S. and J. W. J. Day. 1977. Systems and models: Terms and basic principles. pp. 6–36. *In*: C. A. S. Hall and J. W. J. Day (eds.). Ecosystem Modeling in Theory and Practice. John Wiley & Sons, New York, New York, USA.

IPCC. 2007. Climate Change 2007—The Physical Science Basis. Cambridge University Press, New York, New York, USA.

Keane, R. E., R. A. Loehman and L. M. Holsinger. 2011. The FireBGCv2 landscape fire and succession model: A research simulation platform for exploring fire and vegetation dynamics. General Technical Report RMRS-GTR-255, U.S. Department of Agriculture, Forest Service, Rocky Mountain Research Station, Fort Collins, CO USA.

Loehle, C. and D. LeBlanc. 1996. Model-based assessments of climate change effects on forests: A critical review. Ecological Modelling 90: 1–31.

Mladenoff, D. J. 2004. LANDIS and forest landscape models. Ecological Modelling 180: 7–19.

Reinhardt, E. D., R. E. Keane and J. K. Brown. 2001. Modeling fire effects. International Journal of Wildland Fire 10: 373–380.

Running, S. W. and E. R. Hunt Jr. 1993. Generalization of a forest ecosystem process model for other biomes, BIOME-BGC, and an application for global-scale models. pp. 141–157. Scaling Physiological Processes: Leaf to Globe. Academic Press, Inc.

Scheller, R. M. 2018. The challenges of forest modeling given climate change. Landscape Ecology 33: 1481–1488.

2

Modeling Fundamentals

What You Need to Know to Use This Book

"Excellence is achieved by the mastery of the fundamentals."

Vince Lombardi

A model—a system or thing used as an example to follow or imitate (OED).

ABSTRACT

This chapter presents the general knowledge and terminology needed to (1) understand modeling approaches, (2) complete a successful modeling project, and (3) communicate results with others. First, some fundamental concepts are presented to guide neophyte modelers and to provide a foundation in which to compare and evaluate models. Then, an abridged modeling terminology is provided to understand the book and to create a lexicon for communication with others. Following that is a quick tutorial on the types of models used in natural resource management and the details of their design. And last, the reasons for using ecological modeling for land management and planning along with a description of the model used as examples throughout the book is presented.

What is an Ecosystem Model?

A model is a simplification of reality. More specifically, ecosystem models are computer programs that consist of algorithms that represent a set of important ecological processes to compute ecosystem or landscape dynamics over time. In this book, the term "model" is used in a general sense to encompass all types of computer programs used in ecology for research and management, from a statistical regression equation, to a state-and-transition model, to a complex set of algorithms integrated into a biophysical model that simulates ecosystem interactions. Landscape ecosystem simulation models (LESMs), the primary model type used in examples in this book, are those models that simulate those ecosystem processes over time and space. Models, in a general sense, can be simple, such as a flow diagram, a caricature of a person, or an if-then-else statement, or they can be quite complex, such as a mechanistic ecological process-based model, algorithms that select the most appropriate equation for a collection of regression equation, or a biophysical climate model based on mass and momentum physics. Each model is built for a specific purpose and each purpose is to solve or investigate a specific problem. This chapter presents the general knowledge and terminology needed to understand modeling approaches, complete a successful modeling project, and communicate results with others.

Understanding Modeling for Natural Resource Management

There some fundamental concepts that potential users of ecological models should know before they start a modeling project. Understanding and accepting these views will save countless hours worrying over model project details and will save time interpreting the results. They are:

1. **Models may be the only tool available.** It may be that the lack of comprehensive databases, the limited scope of previous studies, the lack of available expertise, or the advent of the Anthropocene has precluded conventional means of addressing ecosystem response to management actions and that ecological simulation models are the only tool remaining to address management issues.

2. **Models are imperfect representations of the real world.** It is impossible to build an ecosystem or landscape model that simulates everything, and even if it were possible, the computer resources needed to perform such a task would be tremendous. Therefore, all models simplify the world to explore ecosystem behaviors for a specific objective. As a result, all models have limitations, strengths, and weaknesses that should be addressed in the selection of a computer model and in the interpretation of that model's results.

3. **Models reflect the modeler.** Most ecological models and LESMs were built by one person, or sometimes a team of people, for a specific purpose. Modelers deliberately designed their models to answer a unique simulation objective, and as a result, only a small subset of ecosystem processes were included in their model. There are several reasons why the modeler left ecosystem processes out of the simulation design:

    a. Lack of expertise to properly build the algorithms needed to simulate the processes;
    b. Lack of data to parameterize the processes;
    c. Impossibly high levels of complexity in the processes that precluded model simplification;
    d. Inadequate or contradictory literature sources on how to build the algorithms;
    e. Assumption that the missing processes had low importance in completing the modeling objective;
    f. Neglected to include the processes or no knowledge of the processes;
    g. And most commonly, the modeler assumed that the ignored processes were ecologically unimportant.

As a result, it is rare that existing models can fully satisfy the objectives of other modeling objectives without some modification. Again, there will never be a simulation model that is a perfect representation of nature nor one that is appropriate for every objective.

4. **Model results are best interpreted in a relative not an absolute context.** It is rare that any model can provide accurate predictions over a wide range of conditions, especially highly complex models, because of the inherent uncertainty in model construction, initialization, parameterization, and interpretation. However, many models simulate ecosystems with great precision and while the results may not be accurate, the differences across runs or scenarios may be informative and useful. Therefore, the best way to use an ecosystem model is by comparing multiple runs over disparate simulation scenarios. Using a model to predict what will happen at some time in the future is not nearly as reliable as comparing model results across multiple scenarios to understand the factors that contribute to the differences in predications.

5. **Model results should be interpreted within the scope of model assumptions and design.** It is best if all of the assumptions and limitations of the model are known before the actual modeling project is initiated. Then the user can design, prepare, and interpret simulation results in the context of those assumptions. This is important because many users feel that models should provide accurate projections but they forget that models are imperfect. They provide a good "guess" but they rarely provide consistently accurate predictions because of their various limitations. Users should interpret the results within the context of the strengths and weaknesses of the model.

6. **Modeling is both an art and a science.** There are times during a modeling project when important decisions must be made based on educated guesses rather than on an abundance of supporting data or scientific knowledge. A critical parameter, for example, might be unknown for an important species and the user is faced with an important decision—measure it, approximate it, or use another model? An educated guess may be the only alternative for quantification, but it may lead to the introduction of additional uncertainty into model results. However, this additional uncertainty may be

acceptable under the modeling objectives and it can be incorporated into the interpretation of results. Understanding uncertainty and its propagation is critical to designing modeling experiments, but uncertainty is amazingly elusive to quantify. Therefore, modeling and project design is both a creative and a scientific process. Because of this, all creative solutions to the myriad problems encountered during modeling projects must be documented and then taken into account when interpreting the results.

It is important to remember that models are a direct reflection of the modeler, the modeling objective, the resources available for modeling (funding, computers), and the state of the science at the time of model release. Too often I've found that model users are disappointed or highly suspect of model results, and this is often because they did not understand the above principles before they started their modeling project.

Understanding Modeling Terminology

The field of modeling science uses a set of specialized terms to promote efficient communication across modelers and users of models. However, this terminology is rarely consistent across the many ecosystem modeling publications that users will read during their projects. Therefore, a unique set of terms (in bold below) are defined below to be used throughout the book to describe models and modeling activities. These terms will also be important in communicating findings of a modeling project to others.

Uncertainty

This is a quick and dirty short course into understanding uncertainty in terms of modeling science. **Uncertainty** occurs because limited knowledge makes it impossible to accurately predict a future outcome. As model complexity increases to better represent ecological systems, there is a concomitant need to both identify potential sources of uncertainty that result from this complexity and to quantify their impact so that appropriate management options can be evaluated with confidence and in the right context (Ascough et al. 2008). Uncertainty in model predictions is caused by many factors, such as imperfect values for parameters, limitations in model algorithms, inadequate knowledge of the modeler or user, inconsistencies in scale and resolution within the model, and inappropriate initial values (Zimmermann 2000, Jorgensen 2017). And, of course, all of these uncertainties can propagate throughout the model's simulated processes (Beck 1987). In this book, sources of uncertainty are divided into two components - modeling error vs natural variation. Zimmermann (2000) refers to this as subjective impression vs objective fact, respectively. **Modeling error** is the uncertainty resulting from shortcomings of modeling, and it is further divided into two subcomponents: design and application errors. **Design errors** result from faults in model construction or limitations in model design, while **application errors** result from controllable sources when the model was wrongly implemented, such as improper parameterization, inappropriate use of the model, and inadequate initializations. These sources of error are often correctable in some modeling projects but may require abundant resources. **Natural variation**, often called white noise or chaos, is the amount of variation that occurs in nature due to the inherent complexities in ecosystems and is nearly impossible to reduce using better methods or data. This is true ecological variation that must be accounted for in land management actions. Natural variation is not considered model error but rather uncertainty that results from the complexity of ecosystems that precludes representation in models.

A simplistic illustration of the elements of error, accuracy, precision, and natural variation provides a context to understand model uncertainty (Figure 2.1). The smallest possible scatter of points from perfect data represents the natural variation that occurs in the ecosystem for a specific variable regardless of how accurately it is measured. This variation includes uncorrectable measurement errors, natural dynamics of change, and time-space influences. The size of the scatter above the smallest range indicates the consistency of model predictions, often called **precision.** While natural variation is the smallest possible scatter of points, larger point scatters include modeling design and application errors along with natural variation. The average distance of the points to the actual value is called **accuracy.** If a large scatter of points is directly over the actual value (Figure 2.1), then model predictions are considered accurate but not precise (Figure 2.1). And if a small scatter is not centered over the actual value, then the results are considered precise, but not accurate. If the scatter is

16

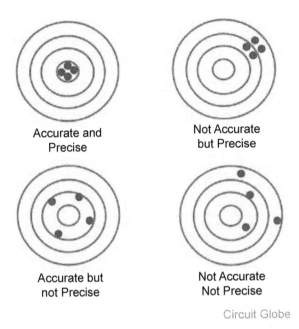

Accurate and Precise

Not Accurate but Precise

Accurate but not Precise

Not Accurate Not Precise

Circuit Globe

Figure 2.1. An illustration of the difference between accuracy and precision to understand uncertainty (from Circuit Globe). If the spread of points is the smallest possible, then the scatter represents a measure of natural variation, but as the scatter widens, model uncertainty in design, construction, and parameterization of the model becomes important.

has a small range, predictions are probably the best the modeler or user can do, albeit it is nearly impossible to quantify the smallest possible scatter. But, if the scatter is wide or imprecise, then the model needs adjustment to fix the controllable errors (see Chapter 6). The entire scatter of points and its relative distance from the actual value is the modeling uncertainty. Complications arise when the "actual" value used as a reference to quantify accuracy and precision has great natural variation and large measurement errors.

The IPCC (2007) uncertainty evaluation guide can be used throughout this book as a way to informally rate various details involved in the modeling tasks (Table 2.1). This is a quick and easy way to rate uncertainty for a parameter's value, a segment of simulated results, or a set of initial conditions. These uncertainty indexes can be entered into spreadsheets

Table 2.1. A classification of uncertainty to use in a modeling project. Level numbers or terms can be used to describe how well a parameter is quantified, for example, or how well an algorithm represents an ecological process. The project team can even rate the model results using this scale based on all uncertainty evaluations.

Level	Term	Description (probability)
1	Virtually certain	99–100% likelihood of the outcome
2	Very likely	90–100% likelihood of the outcome
3	Likely	66–100% likelihood of the outcome
4	Maybe likely	33–660% likelihood of the outcome
5	Unlikely	10–33% likelihood of the outcome
6	Very unlikely	0–10% likelihood of the outcome
7	Extremely unlikely	< 1% likelihood of the outcome

and shared with others when model output is interpreted. For example, the project team may decide that the uncertainty of model results of vegetation projections has a level 2 "Very likely to occur" assignment based on the quality of the initialization, parameterization, and calibration phases. Even though the uncertainty assignments are highly subjective, they do provide some qualitative description of the value of the simulations to the management decision for which they will be used.

Modeling science

There is an entire lexicon on how "models" are scaled up or down to create a particular application (Figure 2.3). All models start from **theory** or research findings; theory and knowledge form the most fundamental layer of all model designs. From that knowledge, **algorithms** can be designed to represent basic ecological processes or fundamental modeling building blocks. The calculation of plant respiration, for example, may involve only one formula with two dynamic variables. Algorithms can be stitched together using computer code to create a **function.** Evapotranspiration, for example, may require over eight separate algorithms and a number of specific parameters to compute evaporation and transpiration (Federer 1975). Functions are then sewn together for a specific purpose to create a **module.** The evapotranspiration function can be integrated with soil water functions to simulate soil moisture. Modules are then combined

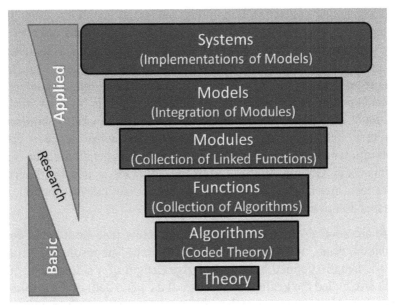

Figure 2.2. The scaling of various ecological models from theory to systems. Theory is first coded into algorithms and these algorithms are integrated into functions that represents ecosystem processes. Functions are then fit together to form modules that simulate a collection of interrelated processes. Modules are then put together to create a model that simulates some ecological entity. Then models are linked to other models to create a system that the user ends up using. An example is when an ecological model is linked to a graphical user interface and a statistical package that contains complex statistical analyses software. Taken from Mark Finney (pers. Communication).

and integrated to create a computer **model.** Using our examples, plant respiration, evapotranspiration, and soil moisture modules can be integrated with modules for photosynthesis to compute stand productivity (Collalti et al. 2014). And last, many computer models are combined along with user-friendly interfaces for data input and output to create a **system.** Systems are often used by managers or scientists when they need many executions of a particular set of models. Systems are usually composed of multiple models and sometimes modules that are sutured together for a specific purpose, and systems also contain user-interfaces for entering input data and summarizing output results.

There are several spatial ecological terms that are often used in modeling science. In forestry, the word "**stand**" is often used to define an area with a homogeneous vegetation conditions; landscape ecologists often refer to this area as a "**patch**" and plant ecologists sometimes refer to this as a "**community**". In digital mapping, the stand, patch, or community is represented by a polygon, a pixel, or a group of adjacent pixels. A **vegetation community** in this book is defined as a unique assemblage of plant species and their relative abundances and it is often an attribute of a stand. The pattern of stands across space is often called **landscape structure** and the various types of these stands across space is called **landscape composition** (e.g., percent of area in vegetation cover types) (Forman and Godron 1986). Landscape structure is usually described by various metrics, such as contagion, patch size and distribution, and interspersion (McGarigal and Marks 1995).

There are several other landscape ecology terms that are used to define resolution, grain, and scale of ecological models. The resolution of raster maps is usually defined by the **pixel size** (length of a side of the square pixel, also called the **grain**) and the area of the pixel defines the **minimum map unit**. Vector maps can be any resolution so usually a minimum mapping unit for a polygon is specified to define resolution. A minimum map unit indirectly relates to **mapscale** in resolution.

Modeling variables

The computer program that represents the model is usually designed to have three types of variables (Swartzman 1979). **State** variables represent various ecological entities, attributes, or characteristics with values that change during the simulation (Figure 2.3). Common state variables are a stand's basal area, carbon, or soil nitrogen. State variables are influenced by various ecosystem flows that cause increases and decreases in the state variables and these flows (**flux** variables) represent ecosystem processes. Examples include water loss from evapotranspiration, biomass consumption by fire, and nitrogen loss via decomposition. The computation of simulated fluxes and how they change state variables almost always require **intermediate** variables, which are variables used to determine transitional products within the algorithm. For example, stomatal conductance is first calculated from weather, soil, and species information, and then it is used to compute evapotranspiration and photosynthesis (Running and Hunt

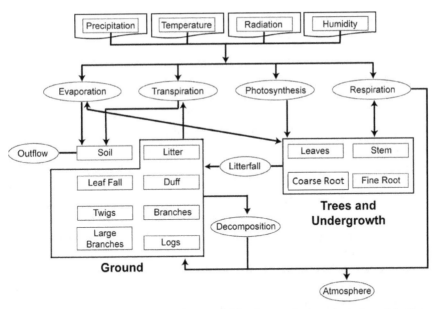

Figure 2.3. An example of the various variables in a typical ecological model. State variables are represented by boxes, the ecological processes (flux variables) are represented by ovals and arrows, and the input variables are represented by the nested boxes. This modeling diagram is for the simulation of stand level processes in the FireBGCv2 model as documented in the Keane et al. (2011) model manual.

1993). And last, there are many types of **output** variables that are written to computer files for later analysis. **Response** variables are those output variables that are used to directly answer the modeling project's objectives and **explanatory** variables are used to explore and explain behaviors of the response variables. **Diagnostic** variables are used to explore model behavior while changing a specific parameter, process, or algorithm in the various phases of a modeling project.

Each modeled flux or process is usually represented by a coded equation that almost always has constant or dynamic **parameters,** which are values that quantify coefficients or threshold values in that algorithm. Values for static parameters are set within the computer code and the user often cannot change these values, but dynamic parameters are usually input into the model by the user (Chapter 5). The computer model often reads values for **input** variables, which can be parameters or initial values. **Initial**

values are the quantification of the starting value of a state, intermediate, or flux variable (Chapter 4).

Modeling tasks

There are a number of other tasks that need to be accomplished before, during, and after a modeling project. First, the modeling project has to be **designed** (Chapter 3). This usually involves the crafting of a comprehensive objective, defining of the project design criteria, selecting of the best model to use to successfully accomplish the simulation objective, defining the simulation landscape, and creating the simulation plan. Model selection is a complex evaluation of multiple factors that many find difficult because of their lack of experience in modeling. The simulation landscape is the area to be simulated; it can be implied by the scale of the model (e.g., stand model) or it is explicitly defined by a digital map (e.g., LESM). The simulation plan is how the model will be used to answer the simulation objective and it usually involves a full factorial design of factors by levels (i.e., management alternatives) where each combination of factor and level is called a **scenario** (e.g., two factors of climate and management where climate has two levels—current and future hot, dry—and management has two levels—no action and thin 1% of the landscape per year).

Once the project is designed, there are six major modeling phases and these form the primary topics (chapters) of this book—initialization, parameterization, calibration, validation, and execution, and analysis (Figure 2.4). In the two beginning modeling phases, **initialization** is the process of quantifying the initial conditions that form the starting place for model simulation (Chapter 4), and **parameterization** is the quantification of the parameters needed by model algorithms (Chapter 5). These two phases (initialization, parameterization) can take place independently and can be accomplished as separate tasks in most modeling projects. They usually involve using the same data whether collecting new or compiling existing data in the field or synthesizing data from the literature. The input data can divided into two classes: data needed to define a starting point in the model (initialization) and data needed to quantify a parameter (parameterization). Each of these phases prepares the model for execution.

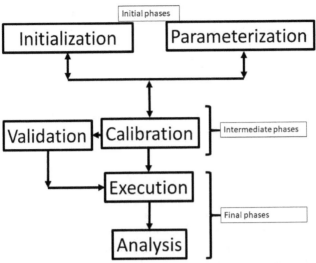

Figure 2.4. The six critical phases of a modeling project. Six chapters in this book (Chapters 4 through 9) contain flow diagrams of each of the steps involved in the implementation of these six phases and the color of each phase is used in the subsequent chapters. Initialization and parameterization are the first, or initial, phases of a modeling project and can be done simultaneously, while the calibration and validation are the intermediate phases. Execution of the model and analysis of the results are the final phases in a modeling project.

In the two intermediate phases, **calibration** is the adjustment of model parameters to increase realism (Chapter 6), and **validation** is the estimation of the accuracy and precision of model results (Chapter 7). Calibration usually occurs after most parameters and initial conditions are quantified. Once the model is calibrated, it is possible that the modeling project it is ready for execution but an estimate of model uncertainty is often needed to evaluate and interpret model results. So, validation is usually done after calibration and is an independent task that indirectly determines model uncertainty. Validation may require that the simulated output be compared against measured data to determine accuracy. A **sensitivity analysis** is sometimes done on complex ecosystem models during the validation phase to determine a parameter's importance to the simulation response variables. This involves changing an important parameter by a pre-determined amount and evaluating the influence of this change on model response variables (see Chapter 7). The **response space** is a multidimensional cloud of data points of important output or response variables that theoretically defines the entire range of model behavior. It is created by incrementally changing each of the important model parameters across a range of values to map the entire space of possible effects in the response variables. Calibration and validation, along with sensitivity analysis, provides several specific knowledge to the model user and team: (1) a notion of the uncertainty and accuracy of model results, (2) confidence in using the model, and (3) experience and knowledge of model behavior important in the interpretation of model results.

Only after the previous four tasks are done, should the model actually be used in a simulation project. In the two final phases, **execution** is the process of actually running the model to complete the simulation design of the modeling project (Chapter 8), and **analysis** is the process of synthesizing and summarizing model results to complete the modeling project and answer the modeling objective (Chapter 9). Analysis usually involves importing the simulated data into a data management system, such as spreadsheets or data bases, and then linking these databases to statistical programs to perform advanced statistical and graphic analysis. Analysis tasks are also done during the previous initial and intermediate modeling phases for a wide variety of purposes; calibration, for instance, might involve the graphical output of all state variables over time with and without fire to evaluate if the model is working correctly (Chapter 6).

Modeling people

There are several definitions used in this book to describe the type of people that are involved in a modeling project. The first is the **modeler** who is a person that built the model or it is an expert on modeling—it is not the person who is using the model for natural resource management applications. The modeler could also be an advanced user and co-developer of a specific model, or the developer of another model. The **user** is a generic term for anyone who uses a model and the primary person for who this book was written. The user could be the modeler, who is using the model for a specific purpose, the advanced practitioner, who has used the model extensively, or the technician that simply pushes the button to run the model. This book assumes that the user is involved in the planning, design and implementation of a modeling project. However, most modeling projects rarely depend on just one person; usually modeling projects are implement by a team of people (**modeling team**). The head of the modeling team is called the **project manager,** and this book assumes that the project manager is the primary user. A **programmer** is a person who programmed the model or programmed revisions and modifications to the model; the programmer can be the modeler or the project manager, or it can be a software engineer employed by the modeler to write code. Often in ecological modeling research projects, the modeler, user, programmer, and project manager are the same person. **Field crews** are usually hired to collect the data needed to initialize, parameterize, validate, and calibrate the model.

Model descriptors

Pairs of contrasting terms are used to describe the design and application of the models presented in this book and they could be used by the modeling team to understand the intricacies of evaluated models (Table 2.2). This list of contrasting pairs is by no means exhaustive, but it does provide a lexicon for efficient communication.

The first, and perhaps the most important, pair is **simple** vs **complex**. This straightforward gradient of complexity is a wonderful way to describe the general construct of many ecological models and provides an important context in which to evaluate and use models (Figure 2.5) (Canham et al. 2004). For example, simple models are easy to learn, use, and understand, while complicated models require abundant training, greater computing resources, longer project times, and more data (Grant and Swannack 2011). Conversely, complex models provide greater exploratory power, extensive investigative output variables, and often the ability to explicitly simulate climate change impacts (Lucash et al. 2018), while simple models have a limited set of output variables and most of the results are highly dependent on input parameters, but they are cheaper to build, require less computer resources, and are less demanding in input parameter requirements (Jorgensen 2017). Complex models often require teams of people to build the various components; need abundant software and hardware resources; and take a long time to quantify all needed parameters (Elizabeth et al.

Table 2.2. A summary of the advantages and disadvantages of the contrasting terms defined in this chapter and used throughout the book.

Terminology	Advantages	Disadvantages
Simple vs	Easy to learn, prepare, and run; uses few computer resources	Limited output; difficult to validate
Complex	Broad range of output variables; novel results; gain valuable ecological insight	Uses abundant computing resources; requires extensive expertise in a wide variety of fields; difficult to understand and use
Stochastic vs	Useful for unstudied or highly complex ecological processes	Demands multiple runs to quantify variability
Deterministic	Same result each time; need only one run; easy to debug and understand	Unavailable for many unstudied or complex ecological process
Empirical vs	Low uncertainty; easily validated; readily available	Can't expand beyond scope of data; climate change and novel management strategies may make past empirical studies questionable
Mechanistic	Incorporate interactions; learn new ecosystem behaviors; integrate climate change	Unknown for many ecosystem processes; complex and difficult for managers to understand and interpret (i.e., black box)
Equilibrium vs	Easy to calibrate; understandable behaviors; evaluates potential	Often precludes disturbance simulation, especially severe disturbances; not entirely realistic in most ecosystems
Non-equilibrium	Explore novel and anticipated responses; more realistically portrays landscapes and ecosystems	Difficult to calibrate; unstable in questionable parameterizations; small changes in parameters may yield large changes in response variables
Spatial vs	Include effects of spatial relationships; stratify results by zone; impose area-based treatments	Difficult to initialize, parameterize, and execute; requires abundant computer resources; results complicated
Non-spatial	Quicker; easier to initialize, parameterize, and understand	Need to run multiple times for each simulated stand; no spatial influences; limited management applications
Vector vs	Scales up or down depending on mapscale; computationally efficient	Difficult to include in simulations; difficult for managers to understand

Table 2.2 contd. ...

... Table 2.2 contd.

Terminology	Advantages	Disadvantages
Raster	Easy to integrate into models; compatible with remotely sensed products	Often results in more computation; inefficient mapping of management units in space
Prognostic vs	Predict what will happen in the future	One-dimensional look at what could happen; doesn't rectify time and space scales
Exploratory	Understand ecosystems and models	Does not generate results management can include in planning documents
Research vs	Explore landscape behaviors	Rarely addresses management needs
Management	Provide projections for alternative management strategies	Rarely provides insight into landscape behaviors

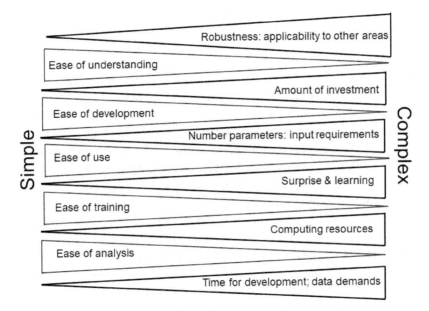

Figure 2.5. The tradeoffs between simple and complex models and the gradient of effort needed for simple to complex models for a number of tasks. The vertical width of the triangle at any given point represents the magnitude of effort relative to the base of the triangle.

2003). The real value of complex models is that they often include the complicated interactions of ecosystem and landscape processes over the appropriate time and space scales (Gustafson 2013). However, the downside of this complexity is that the implementation of the algorithms that represent modeled ecosystem processes are rarely done at the same level of resolution, detail, and accuracy across all processes, and as a result, there is a great deal of uncertainty that propagates across the many interacting variables and algorithms so it is difficult to properly address the behavior and accuracy of the model across all possible situations (Jakeman et al. 2006).

Another important term pair is **stochastic** vs **deterministic**. Stochastic models use probability functions to represent overly complex or poorly understood ecological processes (Black and McKane 2012). Wildland fire ignition, for example, is often represented by a probabilistic model based on fuel, weather, and topography variables because the process of ignition is dictated by the complex interactions of many processes operating over multiples scales of space and time, such as storm tracks, lightning strikes, and fine-scale fuel complex conditions (Fuquay et al. 1979, Balch et al. 2017). Therefore, stochastic models rarely generate the same results after identical simulations. Deterministic models, however, always generate the same results each time the model is executed as long as parameters have not changed (May and Oster 1976). Deterministic models do not have stochastic functions, yet stochastic models may have many deterministic functions. Because of compounding errors and uncertainty, stochastic variation often increases with simulation time (Figure 2.6). The most important thing to remember when using a stochastic model, or a model with stochastic elements, is that it needs to be run multiple times to ensure you have captured most of the variation in modeling results. Most people use 10–20 stochastic model simulations to capture output variability, but this really depends on the model being used and how many stochastic algorithms are included in the model and their influence on model results. If time and computing resources are available, then it is best to perform 50 simulations using the same set of input parameters, and then analyze results for one or more output variables using statistical bootstrap analysis to determine the minimum number of runs needed to capture the envelope of modeling results (Figure 2.6) (Durrett and Levin 1994, Black and McKane

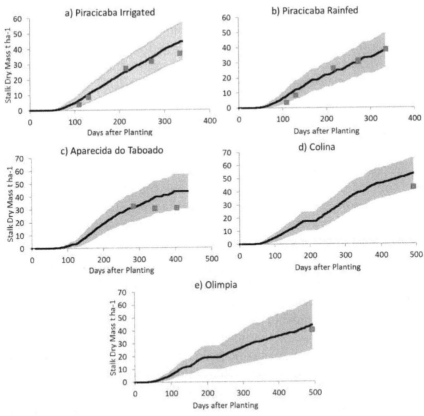

Figure 2.6. A figure showing the range of output data generated from stochastic models from Marin et al. (2017). Here the stalk biomass of five tropical seedling cultivar trees is shown over time. The black line is the daily average of 6000 stochastic simulated samples, the gray area refers to the variation of one standard deviation around the daily mean value, and red squares are the observed data. Note that the stochastic variation increases with time showing increasing and compounding uncertainties in the model.

2012). If time is short, then users should use at least 5–10 simulations and test at the end of the project if enough runs were made (calculate variability across replicates and test against the variability across factors in the simulation experiment).

Models can be either **empirical** vs **mechanistic**. Empirical models are usually those built from actual data using statistical analysis and modeling, such as growth and yield models (Box and Draper 1987). Mechanistic models are those that use basic ecophysiological and physical equations and algorithms to simulate specific ecological processes (Schoener 1986). Many mechanistic models use empirical modeling to parameterize mechanistic relationships, and vice versa. For example, some empirical models use process-based variables as independent variables (e.g., predicting basal area from weather) in regression analyses, while some mechanistic models are parameterized by statistical analysis. Most ecological models, especially LESMs, are built using a complex melding of mechanistic and empirical functions. As a result, these two terms are only used in a general sense—mechanistic models mostly model ecological processes using basic physical principles and empirical models are basically an integration of statistical models generated from field data, such as least square regression equations.

Another dichotomy is between **equilibrium** vs **non-equilibrium** models (DeAngelis and Waterhouse 1987). Models that produce results that appear to level off or converge to a stable specific behavior or result are often called equilibrium models. Non-equilibrium models may not converge to a solution because of the many instabilities and interactions in the modeling design, or there may be multiple behaviors for the same set of inputs because of modeling uncertainty (Keane et al. 2015). It is important to interpret results from each model in different contexts. For example, if trends in a landscape conditions haven't stabilized after some amount of simulation time and the model is an equilibrium model, then the more time must be added to the simulation to obtain equilibrium. And with some non-equilibrium models, the results may diverge in greater magnitude as time increases.

Models can also be **spatial** vs **non-spatial models.** As mentioned, spatial models simulate ecological processes over space (Sklar and Costanza 1991), but not all spatial models explicitly simulate spatial processes (Turner et al. 2001). **Spatially explicit** models simulate spatial processes across the landscape, such as wildlife movement, water flow, and seed dispersal. Sometimes a spatial model is simply a stand-level model where the input and output apply to a specific pixel or polygon and the influence of ecological processes from surrounding pixels and polygons are ignored.

Non-spatial models are ecosystem models that represent a footprint on the ground, but the size, shape, and resolution of that footprint are intrinsic properties of the model.

For spatial models, there are usually two ways to represent entities in space—**vector** vs **raster**. Vector models simulate landscape units as polygons defined by line vectors. Raster models use grids of pixels to define the landscape and each pixel has a value associated with the simulation. Vector spatial models usually have algorithms that define and modify vectors based on complicated topology and this can be somewhat complicated. As a result, many ecological landscape models use raster designs because of the simplified representation. However, vectors are more accurate and often demand less computational and storage resources than raster structures.

Usually, ecological models are used in two ways—**prognostic** vs **exploratory.** Prognostic applications involve predictions of what will happen to the landscape or stand at certain times in the future, while exploratory applications involve understanding responses in landscapes and ecosystems over time from changes in various biophysical factors such as climate, disturbance, vegetation, or management regimes. Prognostic models are commonly used in resource management and a common prognostic application is to use a mensuration model to determine increases in timber volume after 50 years (Stage and Wykoff 1993). Exploratory models are used in a broader sense; for example, to evaluate the changes in fire regimes and forest succession over time (Loehman et al. 2010). This is a subtle but important point. Prognostic models are used to predict what will happen in an absolute sense (e.g., landscape composition in 50 years), while exploratory models are used to evaluate over longer time spans (e.g., how will wildland fire regimes change under warming climates). The real difference between the two really comes down to uncertainty and stochasticity. Highly uncertain or highly stochastic models are rarely used as prognostic tools. An example would be in the assessment of spatially explicit fire regime shifts. Nearly all spatial fire models need to stochastically simulate fire ignition or fire starts because a deterministic approach is difficult at this time (Keane 2012). As a result, fires occur in different places and different times across each model run. Therefore, it is difficult to predict landscape composition after 50 years because it all depends on where and when the fires occurred (Keane et al. 2006). An exploratory approach is the primary use of highly stochastic models. Moreover, if the model has a great deal of uncertainty in the parameter valuation and algorithm construction, then its use should be avoided for prognostic applications unless the inherent uncertainty can be quantified.

One last pair of terms specify the eventual application of the model – **management** vs **research.** A management model was developed specifically for a common land management analysis task, while a research model is usually used to explore various responses to changing parameters for higher level research analyses. Many believe that research models will eventually become management models, but that is not always the case. Most research models are so computationally demanding and so parameter heavy that they will never find a use in management. However, many management models are often used in research studies, but their limited flexibly and simplistic structure often confine model applications to simple comparisons. Some research models are built with no real management application; neutral models, for example, often employ derived hypothetical landscapes to explore landscape behaviors (Gardner and O'Neill 1991, Fall and Fall 1996). Generally, management models are simpler to use, easier to understand, and better designed to answer resource questions. The problem is that the research models often represent the best available scientific knowledge, so many users want to employ the more state-of-the-art program for a specific task because it is more acceptable to the public.

Understanding the Different Types of Models

Most of today's ecological models, especially LESMs, are often built by merging components of many ecological models together into one program. As a result, it is difficult to classify complex models into generalized categories or types (Keane et al. 2004, Scheller and Mladenoff 2007, He 2008). Using the terminology presented previously, the LANDIS II model (Scheller et al. 2007), for example, could be called a complex spatially explicit raster model that uses a mechanistic approach developed for both management and research applications. Albeit, there are many types of models and many types of classifications in the modeling literature that may confuse many people, especially novice users. Instead of classifying models, this section presents a small introduction into the types of models available by important modeling topics by major topic areas.

Scale

Each model is designed for use at specific spatial and temporal scales (Turner et al. 2001) (Figure 1.1). Ecosystem dynamics are influenced by a plethora of interacting factors that act across scales, so ecosystem models must address the various scales of the processes being simulated (Figure 2.7). In natural resource ecological modeling, there are often three

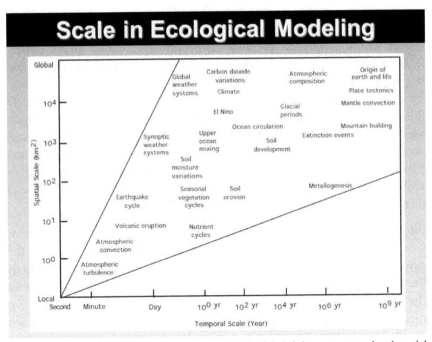

Figure 2.7. A diagram of major ecosystem process and their inherent temporal and spatial scales. Taken from Moritz et al. (2005). Often, ecological simulate the listed processes at finer scales, but results are usually summarized up to the inherent scale of process behavior.

types of models named for the scales at which they operate. The **point** model is a computer program that simulates processes at a specific point or un-specified area on the ground, and the spatial extent of this point is often small (< 25 m²) and intrinsically build into model design. Many fire models, such as FOFEM (Reinhardt and Keane 1998), BEHAVE (Andrews 1986), and BURNUP (Albini 1994), are considered point models because their design is for small (< 100 m²) representative areas. The **stand** or patch model simulates processes at a patch scale (> 100 m²). An example might be a model that simulates stand-level processes such as tree growth in gap phase models (Botkin 1993), timber volume in growth and yield forestry models (Wykoff et al. 1982), or vegetation type change in common state and transition models (Kurz et al. 1999). Inputs to a stand model might be a list of trees within a pre-determined area, and usually, the size of that pre-determined area represents the footprint of the model. As a result, many people call non-spatial models "point" models if the footprint is small or stand models if the footprint is large. The **landscape** model is a spatial model where ecological processes are simulated across many scales. A landscape model, for example, might simulate photosynthesis and respiration at the point level for an individual tree, then summarize these simulations over the trees in each stand on a landscape, and then scale those values up to the landscape level (White 1996). Another important aspect of the landscape model is that they often simulate spatial processes such as hydrology, seed dispersal, and disturbance spread (Scheller and Mladenoff 2007). There are coarser scale ecological models for regional to global applications, such as DGVMs (Purves and Pacala 2008), but these models are rarely used in natural resource management.

Simulation models often have one or more temporal scales. The minimum time it takes a model to go from one time to another time is called the **time step** and the length of time of the simulation is often referred to as the **time span.** As the model is working, the number of years that separate the printing of output is called the **reporting interval.** The time step is a model design criteria, while the time span and reporting interval are often decided by the user. Some landscape models not only have nested spatial scales, but they also have hierarchically nested time scales simulating ecosystem processes at the hour, day, month, and year time steps, such as

34

the FireBGCv2 model (Keane et al. 2011). Ecological models are often classified or described by their inherent temporal scale, or the scale at which the output is most appropriately interpreted (Figure 1.1). **Daily** models, such as weather and hydrology models, often simulate processes at minute or hour time steps, but their results are summarized over days. Daily time steps are usually used in **year, decadal**, and some **century** models, but many landscape models, such as S&T models, are decadal or century models that use a yearly time step. The complex landscape models often mix time steps to simulate ecological processes at the most appropriate temporal scale, but their output is best interpreted in a decadal or century view.

Vegetation models

Vegetation models are those that simulate the development processes of flora in a landscape. These models may range from simulating the growth of individual plants to the seral stage progressions of entire plant communities (i.e., states and transitions). Vegetation is represented by many classifications in management-oriented ecological models including individual plants, cohorts, species, plant functional types (i.e., species guilds), and community types (e.g., cover type). Some models have mixes of all of these attributes. An individual **plant** model has the plant as the finest resolution of simulation in the model, such as a tree, shrub, or grass plant. Each plant is usually represented by a set of attributes, such as biomass, diameter, height, and canopy cover, and the demographic processes of reproduction, regeneration, growth, and mortality of plants may be simulated from multiple ecosystem processes. **Cohort** models simplify the composition and structure of plants in a vegetation community into a specific classification of attributes. Examples of specific cohorts implemented in models are age, plant size, and health classes. Perhaps the most well-known LESM that uses cohorts is LANDIS (Mladenoff and He 1999), which has cohorts of species by diameter classes. **Species** models simulate changes in state variables at the plant species level. For example, phenology in some ecological models is often simulated at the species level rather than the individual plant level. A **plant functional type** (PFT) is often composed of a suite of species that are grouped together for the

35

purpose of simplicity in the simulation architecture (Chapter 5) (Running et al. 1994, Lavorel et al. 1997). Other names for this type of vegetation representation are species functional types (Bonan et al. 2002), species guilds (Wilson 1999), and functional groups (Blondel 2003). In short, PFTs are a great way to group an assemblage of similar species together so that they can be parameterized as a group. Examples of PFTs are needleleaf conifers (used in DGVMs) (Running et al. 1994), fire tolerant shrubs (used in fire effects models) (Thonicke et al. 2001), and C3 vs C4 photosynthetic pathway grasses (used in carbon production models) (Kirschbaum 2004). And perhaps the coarsest vegetation representation is the **community** type which is a generalized classification of broad scale vegetation descriptions that is designed specifically for research and management applications. Cover types, which are often identified as the species with the plurality of some abundance measure (e.g., canopy cover, basal area, density), is the most common classification used to represent forest community types (Eyre 1980), but community types may also be classified by structure as well as composition (Oliver and Larson 1990). Zhu et al. (2006) integrated a cover type classification with a structural stage classification that was based on the height and cover of the canopy. There are coarser representations of vegetation in ecological modeling including biomes (Neilson 1995), ecoregions (Bailey 1995), life and crop zones (Merriam 1898), climate zones (Fovell and Fovell 1993), but these are too broad for most natural resource management analyses. Coarser scale vegetation representations can't facilitate the simulation of vegetation developmental processes in detail because of computing resources, redundancy, high uncertainty, and ecological scale problems (Purves and Pacala 2008).

Several types of common vegetation models are referenced in this book. The simplest vegetation models are often **state and transition (S&T) models** where the development pathways of important vegetation communities (i.e., states) are defined and then time that it takes for each state to transition to another are quantified. Land management has used S&T models for many applications in the past (Barrett 2001, Keane et al. 2006, Wimberly and Kennedy 2008, Barros et al. 2017). Many have described "states" by vegetation composition, structure, age class, and density classes (Keane et al. 1996, Li 2002, Wimberly 2002, Chew et al. 2012). Next in complexity are the **cohort models** that simulate groups of individuals instead of individuals alone (see previous paragraph). For example, LANDIS simulates a number of diameter cohorts (i.e., classes) by species for trees (e.g., 0–5 cm diameter, 5–10 cm diameter, and so on) (Mladenoff and He 1999). And perhaps the most complex are the **individual plant models** that simulate the growth, regeneration, and mortality of individual plants across the stand and landscape. The well-known **gap models**, such as JABOWA (Botkin et al. 1972), SORTIE (Ribbens et al. 1994), and Zelig (Urban 1990), simulate regeneration, growth, and mortality of individual trees at a stand level (Figure 2.8). FireBGCv2 has an embedded gap model

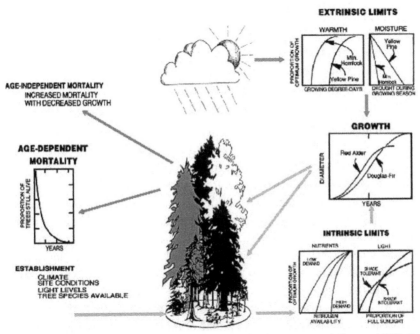

Figure 2.8. A representation of the gap model Forclim (Bugmann and Cramer 1998). In a gap model, a tree grows by computing a maximum diameter and height increment using theoretical mensuration growth curves and those maxima are reduced by various extrinsic (climate represented by degree days and drought curves) and intrinsic (dynamic stand conditions represented by shading and tree density) factors.

simulates individual trees in portions of stands across entire landscapes (Keane et al. 2011).

Disturbance models

Disturbance models and modules simulate effects of various perturbations on ecosystem dynamics. Some may argue that disturbance, not vegetation, is the major factor influencing landscape dynamics over time (Dale et al. 2001, Keane et al. 2013), and therefore disturbances are incredibly important in ecosystem modeling. There are many disturbance models that simulate the dynamics of the disturbance agent (Logan and Bentz 1999, Hatala et al. 2010), but to realistically simulate ecosystem dynamics, those disturbance models must also simulate the interactions of that disturbance on other ecosystem entities, such as vegetation. It is important that the frequency and intensity of the disturbance and the impact of each disturbance event on ecosystem processes and characteristics be represented in a model (Keane et al. 2015).

Disturbances simulations in spatially explicit ecosystem models nearly always involve four separate phases—initiation, spread, effects, and termination (Keane 2013). In **initiation,** the start of a disturbance is often simulated based on climate, topography, and ecosystem conditions. Examples include the start of a wildfire from lightning, the commencement of grazing by ungulates, the start of an outbreak of bark beetles in pine forests, and the beginning of a flood from heavy rain. Once initiated, the disturbance is then **spread** across the landscape. Examples include the spread of fire along vectors of wind, slope, and fuels, the travel of grazing ungulates across a landscape, the flight of bark beetles as they mature and fly to new hosts, and the flow of floodwaters across the basin. As the disturbance is spread in a spatial domain, it impacts the ecosystem in a variety of ways and has various **effects.** Wildland fire, for example, may consume fuel, create smoke, and kill plants or plant parts, while cows may eat specific plant species, bark beetles may cause tree mortality and stress, and floodwaters may deposit nutrients and soil. And lastly, simulated disturbances must end (**termination**). This is perhaps the most difficult of disturbance phases to simulate as it usually involves complex understanding of interacting ecosystem conditions. Limited fuels and wet weather may stop the spread of fire, while humans may round up cattle for slaughter, beetles may kill all the tree hosts, and rains will eventually subside to end flooding.

In this book, disturbance models are often described by the type of disturbance being simulated. Rarely are multiple disturbances simulated by the same module, but many ecosystem models may have multiple disturbance agents and modules. This brings up an interesting dilemma—how many disturbances should be included in a model to realistically simulate ecosystem dynamics (Keane et al. 2015). Often, models contain only those important disturbance agents that create major changes in stand or landscape conditions, such as wildland fire, bark beetles, and spruce budworm in North America, but there are also minor disturbance agents that could create significant impacts if summed across the entire landscape or if there is a severe outbreak, such as dwarf mistletoe, grazing, and endemic diseases. Or, these minor disturbances may become major factors under climate change (Dale et al. 2001). Often modelers will include only the major disturbances in their model and ignore the minor agents. For example, Loehman et al. (2017) used the FireBGCv2 model to assess combined and individual effects of mountain pine beetle, white pine blister rust, and wildland fire on US northern Rocky Mountain landscapes, but FireBGCv2 did not contain the many other disturbance agents that may impact ecosystem dynamics, such as spruce budworm, dwarf mistletoe, and grazing. This does not mean that the model is inaccurate or unrealistic, it only means that the lack of these minor agents in model structure should be assessed in the interpretation of model results.

Climate models

Because of the sheer number and types of climate models, a comprehensive review and synthesis is inappropriate for this book. However, there are some aspects of climate models that users of ecological models should be familiar with to understand and interpret model results. First, climate models are rarely integrated into landscape or ecological models; the outputs from climate models are almost always used as inputs to the ecological model. Climate model output is often stored as daily streams (time series) of one or more of the major weather attributes: temperature, precipitation, humidity, radiation, and wind for example (Friend 1998). Each attribute is quantified using a wide variety of measurements; temperature, for example, is quantified by daily minimum and maximum temperature measured in degrees Celsius, while wind is measured as direction and speed. Historically, ecological models used downloaded weather files from a variety of sources that were reformatted to a native format for climate input. Sometimes, weather is extrapolated across the simulation landscape using a wide variety of computer programs (Hungerford et al. 1989, Thornton et al. 2000). However, raster grids of these measurements are now readily available and seem to be the primary means of representing climate across space in contemporary ecological models (Knutti and Sedlacek 2013, Rupp et al. 2013). These grids can be of varying extents and resolutions from 5° latitude-longitude at monthly time steps to 10 km pixels with daily weather estimates. Some ecological models access these grids directly, while others use specialized software to access a time-space "slice" taken from these grids and format into the ecological model's native format. Many landscape models, for example, use computer programs to extract the daily weather from these large climate grids to summarize to monthly time steps for only the landscape area (Keane and Holsinger 2006). These grids can be historical climate reconstructions or climate forecasts into the future.

In general, there are three scales of climate models that generate the data often used in landscape and ecological models. The **global circulation models** (GCMs) simulate coarse-scale climate dynamics across the earth often using mechanistic approaches. GCMs have broad spatial resolutions (e.g., 0.25 to 1 degree latitude and longitude) but fine temporal resolution (e.g., daily). **Regional** climate models simulate climate at finer scales (< 100 km pixel size) and may use GCM output to set the boundary conditions for the finer scale simulations. Many global and regional models use a mechanistic approach in simulating climate dynamics often employing conservation of mass and momentum approaches into the model. **Local** climate models are used at the finest scales and include both mechanistic and empirical approaches. Local climate grids can be 1 km and finer and include many weather variables at daily time steps.

Users can create climate inputs to ecological models using a bottom up or top down method. In the **bottom up** approach, measured weather variables from weather stations constitute the finest resolution of data available for ecological modeling. These direct observations often include maximum and minimum temperatures, amount of precipitation, humidity, and wind speed and direction, and some weather stations also include cloudiness, soil temperature, and radiation (e.g., short and longwave). These point data can then be extrapolated upwards in scale and over large regions using statistical modeling (Thornton et al. 1997, Thornton et al. 2000). Most FireBGCv2 model applications, for example, often use extrapolated weather station observations across the landscape calculated from the MTCLIM program (Hungerford et al. 1989). Often, weather records from local weather stations have a limited temporal depth (only a few decades of daily data) which poses a problem for long term ecological modeling in that many climate oscillations and teleconnections are poorly represented in a short weather stream. The Pacific Decadal Oscillation, for example, has a multi-decadal variation, while El Nino Southern Oscillations are much shorter (Gershunov and Barnett 1998, Newman et al. 2003). If a weather record does not capture long-term teleconnections, then the temporal variability of that weather record is inadequately represented. One way to expand a short-term weather record is to compute summary statistics for that record and enter them into a weather generator (Nicks et al. 1987, Friend 1998). The weather generator can also be used to create several alternative climate scenarios. However, many ecological models demand multivariate weather records (e.g., temperature, precipitation, radiation) and most weather generators cannot accurately extrapolate synchronized multiple weather variables over time (e.g., the temperature time series is not matched to the precipitation time series). Another method to lengthen weather records is to use empirical methods to statistically model the missing data from a nearby weather station with a longer record.

The climate grids created by GCMs or regional climate models constitute the base data for **top down** approaches (Wilby and Wigley 1997, Salathe et al. 2008). Here, the coarse scale calculations or simulations (e.g., 1 degree by 1 degree) can be directly used as inputs to ecological models, or more appropriately, that coarse data can be scaled down using mechanistic modeling and statistical techniques. Mechanistic climate models can use coarse scale climate as input to simulate finer scale weather records (Pielke et al. 1992). Or, statistical extrapolation techniques can be used to calculate fine scale data (Nabel et al. 2014). Often, finer representations of spatial topography are also included to more accurately simulate weather interactions with aspect, slope, and elevation regardless of approach. The WXFIRE model extrapolated daily DAYMET climate grids (https://daymet.ornl.gov) with 1 km pixel resolution to finer resolutions (30 m) using statistical approaches (Keane and Holsinger 2006).

Other models

There are several classes of ecological models that are different in design and implementation used in land management and deserve mention here (Jørgensen 2008). **The agent-based model (ABMs)** directly simulate autonomous "agents" in a spatiotemporal domain across various scales (Niazi and Hussain 2011, An 2012). These agents can be an individual, such as a beetle, seed, or person, or they can be a collective group (Bone and Altaweel 2014), such as a herd of cows, a clone of trees, or a colony of seals. The goal of ABMs is to explore collective behavior of ecological agents following natural rules rather than as prognostic or predictive implementations. ABMs are also referred to as **individual based models (IBMs)** (Grimm and Railsback 2013) in that the agents are commonly individuals. But IBMs can include a host of other types of individual models, such as those mentioned next under vegetation models. ABMs have been used for many purposes such investigating effects of alternative policies on land use patterns or resting social science concepts analysis (Matthews et al. 2007). **Fuzzy or knowledge-based models** use information technologies (e.g., artificial neural networks, fuzzy logic, artificial intelligence) to simulate ecological systems when little data are available to develop more comprehensive relationships (Chen et al. 2002). **Population models** date back to the Lotka-Volterra model where the focus is on the changes in biotic population structure over time (Law and Morton 1993). These may be IBMs, but more commonly, the population is a state variable and number, age, and health, for example, are attributes of that population.

Reasons to Use Ecological Models

The number of complex interactions across the seemingly countless factors that influence ecosystem dynamics are so great that they boggle the human mind. This is especially true in climate change investigations (Keane et al. 2015). Educated guesses are meaningless in such complicated circumstances, especially for long-term forecasts. This is when people turn to modeling—a way to integrate numerous interactions to explore non-linear, unexpected responses. Therefore, many resource management professionals use ecological models when the prediction or assessment of "real world" ecosystem or landscape responses is so complicated that it is impossible to visualize, evaluate, and understand by the human mind or impossible to measure in the field using empirical approaches (Canham et al. 2004).

Another common reason to use ecological models in natural resource management is to predict future conditions (prognostic application) for evaluation of alternative management strategies (see Figure 1.1). Here, models are executed for several management, disturbance, and/or climate scenarios to determine what the ecosystem or landscape may look like after a certain period of time. In this book, the term prediction, projection, and forecast are used to portray this application. While interesting, this application is not always the best for land management because it often misses some of the rare, larger disturbances events, such as wildfire in North American (Turner and Dale 1998, Scott et al. 2014). Models can simulate planned management activities, but it is incredibly difficult to simulate locations of future disturbances, such as fire and insects, because of high stochasticity of initiation and their large area of influence. Often, most of the natural variability in future landscape predictions comes from the randomness of disturbance initiation (e.g., wildfire starts) (Keane et al. 2011). Unplanned disturbances may overwhelm the subtle differences among scenarios of management alternatives. One can choose to ignore the impact of unplanned disturbances, but then the output becomes somewhat abstract and hypothetical. A better approach is to make the origin and footprint of the disturbance the same in all simulations or use the model for a non-prognostic application.

An alternative to prognostic applications is to describe future long-term changes of landscape of ecosystem dynamics in terms of "regimes" that are often defined as the spatiotemporal expressions of ecological processes (Keane 2013). So, instead of simulating a landscape for 50 years, it would be simulated for 500 years and maybe multiple times. This would provide a comprehensive description of how the vegetation, fire, and hydrologic "regimes" have changed given a specific climate and management futures that, in turn, provides an envelope of possible responses. Changes in fire regimes, for example, would be described by differences in fire rotation or fire frequency, and changes in vegetation would be described using the distributions of community types. This approach simulates ecosystem dynamics over the temporal scales that are more appropriate for evaluating change and, as a result, is used in most modeling projects in my experience.

People also use models when the spatiotemporal domain of empirical data is too limited for a valid analysis. Sometimes, a process model can be used to simulate tree growth in environments where existing in tree ring chronologies are rare (Hunt et al. 1999) or to estimate the intensity and extent of fires in a watershed catchment from paleoecological records (Solomon and Jr. 1984). Growth and yield predictions can be adjusted to account for climate change (Korol et al. 1996, Crookston et al. 2010) and impacts of fuel treatments can be summarized across an entire landscape (Diggins et al. 2010, Chung et al. 2013).

Another common application of modeling in land management is to characterize some expression of ecological variation to use as a reference or benchmark for comparing to other outcomes or conditions. And the most common of these applications is the quantification of an expression of the historical range and variation (HRV) of landscape properties (Keane 2012). Here, the model is initialized and parameterized for historical simulations, such as pre-European fire regimes and historical climate, and then run for hundreds to thousands of years to generate a comprehensive description of the ranges and variations of selected landscape variables, such as vegetation composition and structure. This can then be used as a reference for resilience (Keane et al. 2018), ecosystem health (Landres et al. 1999), or desired future conditions (Blocker et al. 2001).

Landscape Model Used in This Book

The FireBGCv2 model (Keane et al. 2011) is used to provide examples and demonstrate many of the steps, tasks, phases in this book. It is a mechanistic LESM with stochastic elements that simulates basic ecophysiological processes and spatial dynamics to estimate the growth, mortality, and regeneration of individual trees, and shrub and herb guilds. This model was often applied to a specific landscape, the East Fork Bitterroot River (EFBR) basin, a snowmelt-dominated, 105,487-ha watershed (elevations, 1,225–2,887 m) in west-central Montana, USA (Figure 2.9). Annual precipitation in the EFBR averages 41 cm (range, 26–57 cm) with most falling as snow from November to March. The area has primarily a mixed-severity historical fire regime (Arno et al. 2000) with short intervals between low-to-medium intensity fires (mean frequencies of 11–30 years) except in steep terrain, lower-subalpine, and north-facing slopes where stand-replacing fires can occur. It is discussed in detail in several publications (Holsinger et al. 2014, Loehman et al. 2017).

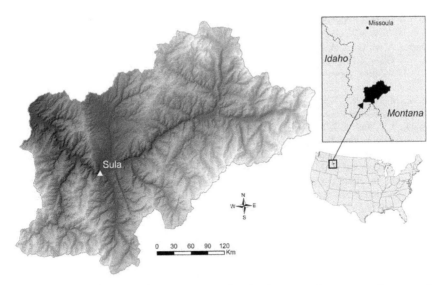

Figure 2.9. The East Fork of the Bitterroot River landscape used as context for some of the examples throughout the book.

References

Albini, F. A. 1994. Program burnup: A simulation model of the burning of large woody natural fuels. Final Report on Research Grant INT-92754-GR, USDA Forest Service, Intermountain Reseach Station, Bozeman, MT.

An, L. 2012. Modeling human decisions in coupled human and natural systems: Review of agent-based models. Ecological Modelling 229: 25–36.

Andrews, P. L. 1986. BEHAVE: Fire behavior prediction and fuel modeling system—BURN subsystem. General Technical Report INT-194, USDA Forest Service.

Arno, S. F., D. J. Parsons and R. E. Keane. 2000. Mixed-severity fire regimes in the northern Rocky Mountains: consequences of fire exclusion and options for the future. Pages 225–232 Wilderness science in a time of change conference, volume 5: wilderness ecosystems, threat, and management, Missoula, Montana, May 23–27, 1999. Fort Collins CO : U.S. Dept. of Agriculture Forest Service Rocky Mountain Research Station 2000.

Ascough, J. C., H. R. Maier, J. K. Ravalico and M. W. Strudley. 2008. Future research challenges for incorporation of uncertainty in environmental and ecological decision-making. Ecological Modelling 219: 383–399.

Bailey, R. G. 1995. Ecoregions map of North America. Miscellaneous Publication Number 1548, USDA Forest Service, Ecosystem Management Analysis Center, Fort Collins, CO.

Balch, J. K., B. A. Bradley, J. T. Abatzoglou, R. C. Nagy, E. J. Fusco and A. L. Mahood. 2017. Human-started wildfires expand the fire niche across the United States. Proceedings of the National Academy of Sciences 114: 2946–2951.

Barrett, T. M. 2001. Models of vegetative change for landscape planning: A comparison of FETM, LANDSUM, SIMPPLLE, and VDDT. General Technical Report RMRS-GTR-76-WWW, USDA Forest Service, Rocky Mountain Research Station, Ogden, UT, USA.

Barros, A. M. G., A. A. Ager, M. A. Day, H. K. Preisler, T. A. Spies, E. White, R. J. Pabst, K. A. Olsen, E. Platt, J. D. Bailey and J. P. Bolte. 2017. Spatiotemporal dynamics of simulated wildfire, forest management, and forest succession in central Oregon, USA. Ecology and Society 22.

Beck, M. B. 1987. Water quality modeling: A review of the analysis of uncertainty. Water Resources Research 23: 1393–1442.

Black, A. J. and A. J. McKane. 2012. Stochastic formulation of ecological models and their applications. Trends in Ecology & Evolution 27: 337–345.

Blocker, L., S. K. Hagle, R. Lasko, R. E. Keane, B. Bollenbacher, B. E. Fox, F. Sampson, R. Gay and C. Manning. 2001. Understanding the connection between historic range of variation, current social values and developing desired conditions. pp. 51–72. *In*: National Silvicultural Workshop. U.S. Department of Agriculture, Forest Service, Rocky Mountain Research Station, Kalispell, MT.

Blondel, J. 2003. Guilds or functional groups: does it matter? Oikos 100: 223–231.

Bonan, G. B., S. Levis, L. Kergoat and K. W. Oleson. 2002. Landscapes as patches of plant functional types: An integrating concept for climate and ecosystem models. Global Biogeochemical Cycles 16.

Bone, C. and M. Altaweel. 2014. Modeling micro-scale ecological processes and emergent patterns of mountain pine beetle epidemics. Ecological Modelling 289: 45–58.

Botkin, D. B. 1993. Forest Dynamics: An Ecological Model. Oxford University Press., New York, NY., USA.

Botkin, D. B., J. F. Janak and J. R. Wallis. 1972. Some ecological consequences of a computer model of forest growth. Journal of Ecology 60: 849–872.

Box, G. E. P. and N. R. Draper. 1987. Empirical model-building and response surfaces. Wiley and Sons, New York, New York USA.

Bugmann, H. and W. Cramer. 1998. Improving the behavior of forest gap models along drought gradients. Forest Ecology and Management 103: 247–263.

Canham, C. D., J. J. Cole and W. K. Lauenroth. 2004. Models in Ecosystem Science. Princeton University Press, Princeton, New Jersey, USA.

Chen, Q., A. Mynett and A. Blauw. 2002. Fuzzy logic and artificial neural network modelling Phaeocystis in the North Sea. pp. 722–728. *In*: Proceedings of Hydroinformatics.

Chew, J. D., K. Moeller and C. Stalling. 2012. SIMPPLLE, version 2.5 user's guide. pp. 363. *In*: F. S. U.S. Dept. of Agriculture (ed.). US Forest Service Rocky Mountain Research Staton, Fort Collins, CO.

Chung, W., G. Jones, K. Krueger, J. Bramel and M. Contreras. 2013. Optimising fuel treatments over time and space. International Journal of Wildland Fire 22: 1118–1133.

Collalti, A., L. Perugini, M. Santini, T. Chiti, A. Nolè, G. Matteucci and R. Valentini. 2014. A process-based model to simulate growth in forests with complex structure: Evaluation and use of 3D-CMCC Forest Ecosystem Model in a deciduous forest in Central Italy. Ecological Modelling 272: 362–378.

Crookston, N. L., G. E. Rehfeldt, G. E. Dixon and A. R. Weiskittel. 2010. Addressing climate change in the forest vegetation simulator to assess impacts on landscape forest dynamics. Forest Ecology and Management 260: 1198–1211.

Dale, V. H., L. A. Joyce, S. McNulty, R. P. Neilson, M. P. Ayres, M. D. Flannigan, P. J. Hanson, L. C. Irland, A. E. Lugo, C. J. Peterson, D. Simberloff, F. J. Swanson, B. J. Stocks and B. Michael Wotton. 2001. Climate Change and Forest Disturbances. BioScience 51: 723–734.

DeAngelis, D. L. and J. C. Waterhouse. 1987. Equilibrium and Nonequilibrium Concepts in Ecological Models. Ecological Monographs 57: 1–21.

Diggins, C., P. Z. Fulé, J. P. Kaye and W. W. Covington. 2010. Future climate affects management strategies for maintaining forest restoration treatments. International Journal of Wildland Fire 19: 903–913.

Durrett, R. and S. A. Levin. 1994. Stochastic spatial models: a user's guide to ecological applications. Transactons of the Royal Society of London B 343: 329–350.

Elizabeth, A. F., D. M. S. Anthony and R. J. Craig. 2003. Effect of complexity on marine ecosystem models. Marine Ecology Progress Series 253: 1–16.

Eyre, F. H. E. 1980. Forest cover types of the United States and Canada. Society of American Foresters, Washington DC., USA.

Fall, A. and J. Fall. 1996. A hierarchical organization of neutral landscape models. *In*: International Association of Landscape Ecology Symposium, Galveston, Texas.

Federer, C. A. 1975. Evapotranspiration. Reviews of Geophysics and Space Physics **13**:442-444.

Forman, R. T. T. and M. Godron. 1986. Landscape Ecology. John Wiley and Sons, New York, NY., USA.

Fovell, R. G. and M.-Y. C. Fovell. 1993. Climate zones of the conterminous United States Defined Using Cluster Analysis. Journal of Climate 6: 2103–2135.

Friend, A. D. 1998. Parameterisation of a global daily weather generator for terrestrial ecosystem modelling. Ecological Modelling 109: 121–140.

Fuquay, D. M., Robert G. Baughman and D. J. Latham. 1979. A model for predicting lightning-fire ignition in wildland fuels. Research Paper INT-217, USDA Forest Service.

Gardner, R. H. and R. V. O'Neill. 1991. Pattern, process, and predictability: The use of neutral models for landscape analysis. pp. 289–307. *In*: M. G. Turner and R. H. Gardner (eds.). Quantitative Methods in Landscape Ecology. Springer-Verlag, New York.

Gershunov, A. and T. P. Barnett. 1998. Interdecadal Modulation of ENSO Teleconnections. Bulletin of the American Meteorological Society 79: 2715–2725.

Grant, W. E. and T. M. Swannack. 2011. Ecological Modeling: A Common-Sense Approach to Theory and Practice. John Wiley & Sons.

Grimm, V. and S. F. Railsback. 2013. Individual-based modeling and ecology. Princeton university press.

Gustafson, E. 2013. When relationships estimated in the past cannot be used to predict the future: Using mechanistic models to predict landscape ecological dynamics in a changing world. Landscape Ecology 28: 1429–1437.

Hatala, J. A., M. C. Dietze, R. L. Crabtree, K. Kendall, D. Six and P. R. Moorcroft. 2010. An ecosystem-scale model for the spread of a host-specific forest pathogen in the Greater Yellowstone Ecosystem. Ecological Applications 21: 1138–1153.

He, H. S. 2008. Forest landscape models, definition, characterization, and classification. Forest Ecology and Management 254: 484–498.

Holsinger, L., R. E. Keane, D. J. Isaak, L. Eby and M. K. Young. 2014. Relative effects of climate change and wildfires on stream temperatures: A simulation modeling approach in a Rocky Mountain watershed. Climatic Change 124: 191–206.

Hungerford, R. D., R. R. Nemani, S. W. Running and J. C. Coughlan. 1989. MTCLIM: A mountain microclimate simulation model. Research Paper INT-414, USDA Forest Service, Intermountain Research Station, Ogden, UT.

Hunt, E. R. J., M. B. Lavigne and S. E. Franklin. 1999. Factors controlling th edecline of net primary production with stand age for balsam fir in Newfoundland assessed usnign an ecosystem simulation model. Ecological Modelling 122: 151–164.

IPCC. 2007. Climate Change 2007—The Physical Science Basis. Cambridge University Press, New York, New York, USA.

Jakeman, A. J., R. A. Letcher and J. P. Norton. 2006. Ten iterative steps in development and evaluation of environmental models. Environmental Modelling & Software 21: 602–614.

Jorgensen, S. E. 2017. Handbook of environmental and ecological modeling. CRC Press.

Jørgensen, S. E. 2008. Overview of the model types available for development of ecological models. Ecological Modelling 215: 3–9.

Keane, R. E. 2012. Creating historical range of variation (HRV) time series using landscape modeling: overview and issues. pp. 113–128. *In*: J. A. Wiens, G. D. Hayward, H. S. Stafford and C. Giffen (eds.). Historical Environmental Variation in Conservation and Natural Resource Management. John Wiley and Sons, Hoboken, New Jersey.

Keane, R. E. 2013. Disturbance Regimes and the Historical Range of Variation in Terrestrial Ecosystems. pp. 568–581. *In*: A. L. Editor-in-Chief: Simon (ed.). Encyclopedia of Biodiversity (Second Edition). Academic Press, Waltham.

Keane, R. E., G. Cary, I. D. Davies, M. D. Flannigan, R. H. Gardner, S. Lavorel, J. M. Lennihan, C. Li and T. S. Rupp. 2004. A classification of landscape fire succession models: Spatially explicit models of fire and vegetation dynamic. Ecological Modelling 256: 3–27.

Keane, R. E., G. J. Cary, M. D. Flannigan, R. A. Parsons, I. D. Davies, K. J. King, C. Li, R. A. Bradstock and M. Gill. 2013. Exploring the role of fire, succession, climate, and weather on landscape dynamics using comparative modeling. Ecological Modelling 266: 172–186.

Keane, R. E. and L. Holsinger. 2006. Simulating biophysical environment for gradient modeling and ecosystem mapping using the WXFIRE program: Model documentation and application. Research Paper RMRS-GTR-168CD, USDA Forest Service Rocky Mountain Research Station, Fort Collins, Co, USA.

Keane, R. E., L. Holsinger and S. Pratt. 2006. Simulating historical landscape dynamics using the landscape fire succession model LANDSUM version 4.0. General Technical Report RMRS-GTR-171CD, USDA Forest Service Rocky Mountain Research Station, Fort Collins, CO USA.

Keane, R. E., R. A. Loehman and L. M. Holsinger. 2011. The FireBGCv2 landscape fire and succession model: a research simulation platform for exploring fire and vegetation dynamics. General Technical Report RMRS-GTR-255, U.S. Department of Agriculture, Forest Service, Rocky Mountain Research Station, Fort Collins, CO USA.

Keane, R. E., R. A. Loehman, L. M. Holsinger, D. A. Falk, P. Higuera, S. M. Hood and P. F. Hessburg. 2018. Use of landscape simulation modeling to quantify resilience for ecological applications. Ecosphere 9: e02414.

Keane, R. E., D. G. Long, J. P. Menakis, W. J. Hann and C. D. Bevins. 1996. Simulating coarse-scale vegetation dynamics using the Columbia River Basin succession model: CRBSUM. RMRS-GTR-340, U.S. Dept. of Agriculture Forest Service Intermountain Research Station, Ogden, UT.

Keane, R. E., D. McKenzie, D. A. Falk, E. A. H. Smithwick, C. Miller and L.-K. B. Kellogg. 2015. Representing climate, disturbance, and vegetation interactions in landscape models. Ecological Modelling 309-310: 33–47.

Kirschbaum, M. U. F. 2004. Direct and indirect climate change effects on photosynthesis and transpiration. Plant Biology 6: 242–253.

Knutti, R. and J. Sedlacek. 2013. Robustness and uncertainties in the new CMIP5 climate model projections. Nature Clim. Change 3: 369–373.

Korol, R. L., K. S. Milner and S. W. Running. 1996. Testing a mechanistic model for predicting stand and tree growth. Forest Science 42: 139–153.

Kurz, W. A., S. J. Beukema, J. Merzenich, M. Arbaugh and S. Schilling. 1999. Long-range modeling of stochastic disturbances and management treatments using VDDT and TELSA. pp. 349–355. *In*: Society of American Foresters 1999 National Convention. Society of American Foresters, Portland, Oregon, USA.

Landres, P. B., Penelope Morgan and F. J. Swanson. 1999. Overview and use of natural variability concepts in managing ecological systems. Ecological Applications 9: 1179–1188.

Lavorel, S., S. McIntyre, J. Landsberg and T. D. A. Forbes. 1997. Plant functional classifications: from general groups to specific groups based on response to disturbance. Trends in Ecology & Evolution 12: 474–478.

Law, R. and R. D. Morton. 1993. Alternative permanent states of ecological communities. Ecology 74: 1347–1361.

Li, C. 2002. Estimation of fire frequency and fire cycle: A computational perspective. Ecological Modelling 145.

Loehman, R. A., J. A. Clark and R. E. Keane. 2010. Modeling effects of climate change and fire management on western white pine (Pinus monticola) in the northern Rocky Mountains, USA. Forests 2.

Loehman, R. A., R. E. Keane, L. M. Holsinger and Z. Wu. 2017. Interactions of landscape disturbances and climate change dictate ecological pattern and process: Spatial modeling of wildfire, insect, and disease dynamics under future climates. Landscape Ecology 32: 1447–1459.

Logan, J. A. and B. J. Bentz. 1999. Model analysis of mountain pine beetle (Coleoptera Scolytidae) seasonality. Enviromental Entomology 29: 924–934.

Lucash, M. S., R. M. Scheller, B. R. Sturtevant, E. J. Gustafson, A. M. Kretchun and J. R. Foster. 2018. More than the sum of its parts: How disturbance interactions shape forest dynamics under climate change. Ecosphere 9: e02293.

Marin, F., J. W. Jones and K. J. Boote. 2017. A Stochastic Method for Crop Models: Including Uncertainty in a Sugarcane Model. Agronomy Journal 109: 483–495.

Matthews, R. B., N. G. Gilbert, A. Roach, J. G. Polhill and N. M. Gotts. 2007. Agent-based land-use models: A review of applications. Landscape Ecology 22: 1447–1459.

May, R. M. and G. F. Oster. 1976. Bifurcations and Dynamic Complexity in Simple Ecological Models. The American Naturalist 110: 573–599.

McGarigal, K. and B. J. Marks. 1995. FRAGSTATS: Spatial pattern analysis program for quantifying landscape structure. General Technical Report PNW-GTR-351, USDA Forest Service.

Merriam, C. H. 1898. Life Zones and Crop Zones of the United States. US Government Printing Office.

Mladenoff, D. J. and H. S. He. 1999. Design, behavior and application of LANDIS, an object-oriented model of forest landscape disturbance and succession. pp. 125–162. *In*: D. J. Mladenoff and W. L. Baker (eds.). Spatial Modeling of Forest Landscape Change: Approaches and Applications. Cambridge University Press, Cambridge, United Kingdom.

Moritz, M. A., M. E. Morais, L. A. Summerell, J. M. Carlson and J. Doyle. 2005. Wildfires, complexity, and highly optimized tolerance. Proceedings of the National Academy of Sciences of the United States of America 102: 17912–17917.

Nabel, J. E. M. S., J. W. Kirchner, N. Zurbriggen, F. Kienast and H. Lischke. 2014. Extrapolation methods for climate time series revisited—Spatial correlations in climatic fluctuations influence simulated tree species abundance and migration. Ecological Complexity.

Neilson, R. P. 1995. A model for predicting continental-scale vegetation distribution and water balance. Ecological Applications 5: 362–385.

Newman, M., G. P. Compo and M. A. Alexander. 2003. ENSO-Forced Variability of the Pacific Decadal Oscillation. Journal of Climate 16: 3853–3857.

Niazi, M. and A. Hussain. 2011. Agent-based computing from multi-agent systems to agent-based models: A visual survey. Scientometrics 89: 479.

Nicks, A. D., J. R. Williams and C. W. Richardson. 1987. Generating climatic data for a water erosion prediction model. pp. 002–011. *In*: Proceedings of the 1987 International Winter Meeting of the American Society of Agricultural Engineers. ASAE, Chicago, IL.

Oliver, C. D. and B. C. Larson. 1990. Forest Stand Dynamics. McGraw Hill, New York, USA.

Pielke, R. A., W.R. Cotton, R.L. Walko, C.J. Tremback, M.E. Nicholls, M.D. Moran, D.A. Wesley, T.J. Lee and J. H. Copland. 1992. A comprehensive meteorological modeling system—RAMS. Meteorology and Atmospheric Physics 49: 69–91.

Purves, D. and S. Pacala. 2008. Predictive Models of Forest Dynamics. Science 320: 1452–1453.

Reinhardt, E. and R. E. Keane. 1998. FOFEM—a First Order Fire Effects Model. Fire Management Notes 58: 25–28.

Ribbens, E., J. A. Silander Jr. and S. W. Pacala. 1994. Seedling recruitment in forests: Calibrating models to predict patterns of tree seedling dispersion. Ecology 75: 1794–1806.

Running, S. W. and E. R. Hunt. 1993. Generalization of a forest ecosystem process model for other biomes, BIOME-BGC, and an application for global-scale models. pp. 141–157. Scaling Physiological Processes: Leaf to Globe. Academic Press, Inc.

Running, S. W., T. R. Loveland and L. L. Pierce. 1994. A vegetation classification logic based on remote sensing for use in global biogeochemical models. Ambio 23: 77–81.

Rupp, D. E., J. T. Abatzoglou, K. C. Hegewisch and P. W. Mote. 2013. Evaluation of CMIP5 20th century climate simulations for the Pacific Northwest USA. Journal of Geophysical Research: Atmospheres 118: 10,884–810,906.

Salathe, E. P., P. W. Mote and M. W. Wiley. 2008. Considerations for selecting downscaling methods for integrated assessments of climate change impacts. Intl. J. of Climatology 27: 1611–1621.

Scheller, R. and D. Mladenoff. 2007. An ecological classification of forest landscape simulation models: tools and strategies for understanding broad-scale forested ecosystems. Landscape Ecology 22: 491–505.

Scheller, R. M., J. B. Domingo, B. R. Sturtevant, J. S. Williams, A. Rudy, E. J. Gustafson and D. L. Mladenoff. 2007. Design, development, and application of LANDIS-II, a spatial landscape simulation model with flexible temporal and spatial resolution. Ecol. Model 201.

Schoener, T. W. 1986. Mechanistic Approaches to Community Ecology: A New Reductionism1. American Zoologist 26: 81–106.

Scott, A. C., D. M. J. S. Bowman, W. J. Bond, S. J. Pyne and M. E. Alexander (eds.). 2014. Fire on Earth: An Introduction. John Wiley and Sons Ltd., Chichester, England.

Sklar, F. H. and R. Costanza. 1991. The development of dynamic spatial models for landscape ecology: A review and prognosis. pp. 239–288. *In*: M. G. Turner and R. H. Gardner (eds.). Quantitative Methods in Landscape Ecology. Springer-Verlag, New York.

Solomon, A. M. and H. H. S. Jr. 1984. Integrating forest stand simulations with paleoecological records to examin long-term forest dynamics. pp. 333–356. *In*: G. I. Agren (ed.). State and Cahnge of Forest Ecosystems. Swedish Univ. Agr. Sci., Uppsala, Sweden.

Stage, A. R. and W. R. Wykoff. 1993. Calibrating a model of stochastic effects on diameter increment for individual-tree simulations of stand dynamics. Forest Science 39: 692–705.

Swartzman, G. L. 1979. Simulation modeling of material and energy flow through an ecosystem: Methods and documentation. Ecological Modelling 7: 55–81.

Thonicke, K., S. Venevsky, S. Sitch and W. Cramer. 2001. The role of fire disturbance for global vegetation dynamics: coupling fire into a Dynamic Global Vegetation Model. Global Ecology and Biogeography 10: 661–677.

Thornton, P., E., Steven W. Running and M. A. White. 1997. Generating surfaces of daily meteorological variables over large regions of complex terrain. Journal of Hydrology 190: 214–251.

Thornton, P. E., H. Hasenauer and M. A. White. 2000. Simultaneous estimation of daily solar radiation and humidity from observed temperature and precipitation: An application over complex terrain in Austria. Agricultural and Forest Meteorology 104: 255–271.

Turner, M. G. and V. H. Dale. 1998. Comparing Large, Infrequent Disturbances: What Have We Learned? Ecosystems 1: 493–496.

Turner, M. G., R. H. Gardner and R. V. O'Neill. 2001. Landscape ecology in theory and practice. Springer-Verlag, New York, New York, USA.

Urban, D. L. 1990. A versatile model to simulate forest pattern: A users guide to ZELIG version 1.0., Environmental Sciences Department, The University of Virginia, Charlottesville, Virginia, USA.

White, J. D. 1996. Spatial, and temporal scale effects on assessment of a regional ecosystem model: Modeling climate change in Glacier National Park, USA. Ph.D. Dissertation. University of Montana, Missoula, MT., USA.

Wilby, R. L. and T. M. L. Wigley. 1997. Downscaling general circulation model output: a review of methods and limitations. Progress in Physical Geography 21: 530–548.

Wilson, J. B. 1999. Guilds, functional types and ecological groups. Oikos: 507–522.

Wimberly, M. C. 2002. Spatial simulation of historical landscape patterns in coastal forests of the Pacific Northwest. Canadian Journal of Forest Research 32: 1316–1328.

Wimberly, M. C. and R. S. H. Kennedy. 2008. Spatially explicit modeling of mixed-severity fire regimes and landscape dynamics. Forest Ecology and Management 254: 511–523.

Wykoff, W. R., N.L. Crookston and A. R. Stage. 1982. User's guide to the stand prognosis model. General Technical Report INT-133, USDA Forest Service.

Zhu, Z., J. Vogelmann, D. Ohlen, J. Kost, S. Chen, B. Tolk and M. G. Rollins. 2006. Mapping existing vegetation composition and structure. General Technical Report RMRS-GTR-175, USDA Forest Service Rocky Mountain Research Station, Fort Collins, CO USA.

Zimmermann, H. J. 2000. An application-oriented view of modeling uncertainty. European Journal of Operational Research 122: 190–198.

3

Project Design
How to Plan a Modeling Project

"All problems are solved with good design."

Stephen Gardiner

Modeling project—direct application of an ecological or landscape model for research or management purposes.

ABSTRACT

Designing a modeling project is covered in this chapter through a series of steps that starts with setting the modeling objective and ends with implementing a successful modeling project. Once an objective or set of objectives are decided, the user needs to decide on a set of design criteria that will set the boundaries of various decisions made during the modeling project. These criteria include cost, data, expertise, and computer limitations, along with target accuracies and spatial and temporal resolutions of the project. These criteria are then used to select the right model for the project. Once the model is selected, then the simulation landscape or spatial domain of the project can be delineated. Finally, the actual design of the simulation scenarios is covered, especially the use of hierarchically nested scenarios to simulate ranges of management alternatives and environmental uncertainty. This is followed by various suggestions on how to ensure the design is properly implemented in the project.

Introduction

Hopefully, the reader now has a cursory understanding of modeling science and it's time to start designing a modeling project. This can be as easy as applying an existing model that has been previously parameterized and initialized to a new situation, or as complex as modifying an existing model to be applied to a landscape and ecosystem that has never been studied. This chapter deals with the steps needed to design a modeling project (Figure 3.1), while the remaining chapters deal with the actual

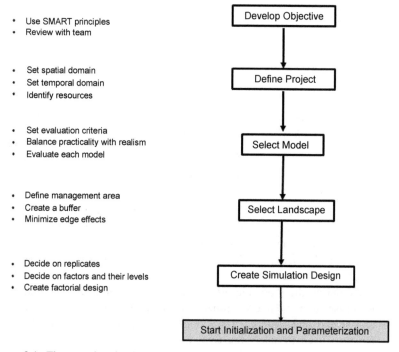

Figure 3.1. The steps involved in model project design. Suggestions on how to complete the steps is summarized on the left side of the figure.

implementation of that project for a natural resource management application.

There are several tasks that must be completed to set a strong foundation and to design the simulation architecture of the project so that all decisions and activities concerning that project have a comprehensive context (Figure 3.1). All of the project steps mentioned in this chapter are sequential and build on each other. First and most important, a succinct objective must be stated so that a solid decision framework is created. Then the limiting factors of the project must be defined, such as available resources, timelines, expertise, and project area. A model is then selected based on those criteria. Once the model is chosen, the temporal and spatial domain of the simulation area can be decided. Last, the simulation project must be designed by deciding on the simulation parameters for model execution including scenario design, simulation time, and output reporting interval. Once these steps are completed, the initialization and parameterization steps described in the next two chapters can be started.

Developing a Modeling Objective

Designing the modeling objective is easily the most important aspect in any modeling project. It is more important than model selection, parameter quantification, or computing resources. Setting the modeling objective sets the context for the entire modeling project and assures that model results will in fact meet decision-support needs. All modeling decisions made during the project will be done by referencing the objective. The one thing to remember from this chapter is to spend the time to craft a useful and comprehensive modeling objective.

While there are many guides on how to construct a practical modeling objective (Keeney 2007), the SMART principles are perhaps the best guidelines for setting objectives in the modeling environment. SMART stands for Specific, Measureable, Achievable, Relevant, and Time-based (Lutes et al. 2006, Bovend'Eerdt et al. 2009). Objectives should avoid generalities and be *specific* about what is needed. An objective of "Simulate fire in northern Australia" is missing key specifics of: (1) where in northern Australia, (2) what attributes will be used to describe fire (e.g., fire intensity, severity, and pattern), and (3) what time period should be represented or simulated. Once the specifics are set, then it is important that the specified objective be *measureable* so that model results can be compared against values reported in the literature and from local studies (see Chapter 7). An objective that states "Compare impacts on forest growth" should include specifics on the variables that will be used to measure forest growth, such as basal area, biomass, or tree diameter (i.e., attributes that are easily measured). It is also important that the simulated variables link directly to the natural resource management planning task. Obviously, all objectives must be *achievable* and *relevant,* but in this book, objectives must link directly to the model being used so that the desired variables are output by the selected model. Many projects have chosen models whose outputs were not useful in management planning and were nearly impossible to measure in the field, such as belowground carbon. And last, the objective must include a *timeline*, both for the project itself and also for the modeling context. For example, the objective might say "Compare impacts … across five decades of simulation and present results at the March 2023 conference". I've found many users make the mistakes of: (1) using goals instead of defining objectives (not specific), (2) failing to mention what is being simulated and what variables should be output (hard to measure), (3) specifying too many tasks for a modeling effort (not achievable), (4) including aspects that are unrelated to the simulation effort (irrelevant), and (5) forgetting to add deadlines and scheduling concerns (not time-based). Without doubt, a well-stated objective is the keystone of a successful simulation design and the foundation of a successful project.

It is strongly recommended that, once the modeling objective has been clearly articulated, the structure and format of the anticipated simulation results should be developed. This may sound odd—designing the tables and figures for a report before the model has actually been selected or run— but it will save time and resources by eliminating needless exploration of inappropriate models and unnecessary model analysis. There is nothing more frustrating than completing months of simulation only to find that the most appropriate variables for answering the objective were never available or output. Or conversely, a model might provide copious output for numerous variables, but when it comes down to writing the report, the user realizes that only a fraction of the terabytes of output was actually needed. It is suggested that the modeler mock up the design of 3–5 figures or tables that could be used to efficiently evaluate and complete the model objective. These tables can be summaries of response variables that represent the ecosystem processes, or syntheses of the relationship of the response variables to the factors (i.e., explanatory variables) that influence them. Think of these two kinds of variables as answers to the questions "What" (what happened=response) and "Why" (why did I get these results=explanatory). Decision theory uses the notion of "consequence tables" as a more in-depth application of the mock tables and figures presented here (Gregory et al. 2012). In consequence tables, the variables and their units are selected and their anticipated magnitude and direction are estimated by each management alternative. While the reader is encouraged to create more tables and figures to explore and interpret model results during the analysis phase (Chapter 9), the primary set of finished figures and tables will ensure the objective is answered.

Last, it is also recommended that modeling projects take a scenario-based approach to project design because of the high uncertainty in model design and implementation. Most models are so complex and demand such extensive parameterizations that accurate results are difficult to obtain. As a result, it is best that the model results be interpreted in a comparative sense rather than using modeling results as the absolute answer. The strengths of ecological models are best realized through comparison of alternatives. Another words, instead of simulating just one management goal or alternative, it may be best to create a set of scenarios that bound the range of possible responses given the modeling objective. For example, instead of simulating the future fire regime for a landscape using one climate change scenario, it is recommended that several nested scenarios be employed, such as three climate scenarios, three fire management scenarios, and three land use scenarios for a total of 27 different scenarios. This will provide a range of possible responses over all possible management alternatives and will give insight into model sensitivity and behavior. Moreover, it is suggested that the scenario results be analyzed across at least two time frames – the management horizon and ecological horizon. Often, the management horizon is one to five decades in the future, encompassing the time scales of planning. However, ecosystem response often can't be evaluated with in that short of time span because many simulated climate, disturbance, and vegetation regimes have yet to cause discernible impacts because of system inertia. Therefore, simulations of one to ten centuries may be necessary to see the cumulative ecological impacts of the simulated entities.

Defining the Project

Once the modeling objective is clearly stated, the overall design of the modeling project can begin. There is a wide array of factors that must be addressed so that the six major modeling tasks in this book can be accomplished with the greatest efficiency. These elements will set the sideboards on various decisions that will need to be made throughout the project development and implementation. For example, these criteria may be used as a reference when deciding the timeline, detail, accuracy, and thoroughness for successful completion of each of the six phases. These criteria mirror the selection criteria for deciding on the most appropriate model discussed in the following section, but the project criteria pertain, not to individual models, but to the resources available to successfully complete the modeling objective.

When does the project need to be completed?

The first specification for the project is the amount of time available for completion. I've found that the timeline is often the most important design element in a modeling project, and this includes report writing and publication of results. Modeling projects that must be completed quickly (months) will probably require simple models that are easy to use because the analyses will be uncomplicated and the interpretation of results straightforward with little exploration and insight (see Table 2.1). Complex modeling projects that use mechanistic models and complicated scenario designs will take a great deal of time and modeling project managers should plan for prolonged analysis. Many modeling projects have been abandoned because people spent too much time in the initialization and parameterization phases, and not nearly enough time in analysis and report writing because time spent in later phases (Figures 1.2, 2.4) is often compressed. Deciding on a "drop-dead" date for the project's end is always best, and some have established three deadline dates—best case (soonest), most likely, and drop-dead (latest). It is also advisable to set deadlines for each of the phases in the project based on the timeline of the project.

Are sufficient computing resources available?

The second specification is the amount of the computing resources available to do the project. Too often, simulation designs are so complex that they can't be implemented on the available computers, or the generated output requires some much disk space that it can't be stored on native disk drives, or output files are so big that it can't be analyzed using conventional statistical programs. Limited computing resources may mean that simple models are the only alternative.

Both hardware and software demands should be considered. Not only does there need to be sufficient computing power to finish the simulations, there also needs to be the software available to perform all of the modeling tasks, such as GIS software for initialization tasks, graphical programs for calibration, and statistical packages for analysis. Often, computing power is measured in two ways—number and speed of processors and amount of resident memory (RAM). The most common problem that I've encountered is the lack of sufficient computer memory for running complex models and there often isn't an informative error message when this happens. Hopefully, model developers have memory recommendations for their programs, but many times, modeling projects find hardware computing limitations by trial and error.

Who is available to help?

The third criteria is the identification of expertise available to help with modeling and analysis chores. It is highly recommended that the developer of the selected model be available for consultations, otherwise, it will be quite difficult to efficiently prepare, run, and evaluate model results. If the developer is unavailable, then it is critical that a relationship be started with other ecological modelers or someone who has used the selected model in the past. Countless modeling projects have spent inordinately long times on various modeling phases that could have been greatly shortened if they had contacted the developer or a modeling expert. It is important that the project manager or primary user get firm commitments from the team and the modeling expert to lend a hand.

There is also a need for expertise for all phases of a modeling project. People must be available to help with installing the model, initializing and parameterizing the model, running the model, dealing with modeling software and hardware issues, and analyzing the results. I have seen people decide to use a complex spatial ecosystem model only to find that they have no idea of how to analyze and interpret the results or how to install it on their computer. It is important that the modeling team be constructed to best facilitate the successful completion of the objectives.

Another important person to have on the project is someone who knows the ecology of the area, especially disturbance regimes, vegetation dynamics, and climate interactions. Many projects were stymied in the calibration phase because there were few available experts to evaluate if the results seem realistic. Another important person to have on the team is a computer specialist; it is critical that someone is around to deal with the common minor to major software and hardware issues that crop up in modeling projects. I've also found that a statistician is critical for helping design valid statistical analyses that evaluate significance because results from many landscape models have both spatial and temporal autocorrelation.

What is the resolution of the project's decision space?

Another criteria is the level of resolution that is needed to make the management decision, which should be able to be inferred from the modeling objective. If a "yes" or "no" decision is all that is needed, then a simple, parsimonious model may be appropriate thereby saving time and effort and requiring little expertise. However, if the decision demands a finer level of resolution of detail that must be quantified across a gradient of scenarios using abundant variables, then more complex models are indicated. Basal area, timber volume, and tree density, for example, might need to be evaluated for three climate scenarios over 100 years of simulation. A well-articulated scale of management decision-making can save time and energy in other project phases. In one modeling project, I was asked to implement a complex S&T modeling effort simulating over 100 vegetation types to develop future scenarios across the Interior Columbia River Basin (Keane et al. 1996), but when it was time to develop management plans, managers decided that the complicated results had to be summarized by grouping the 100 to only nine vegetation types for simplicity (Hann et al. 1997). We could have drastically decreased the time needed for model initialization and parameterization had we only modeled nine instead of 100 types.

Spatial resolutions are perhaps the most important criteria in landscape modeling projects. Finer spatial resolution landscapes (e.g., < 90 m pixel sizes) are particularly demanding in terms of computer resources and simulation times, while coarser resolutions usually result in greater uncertainty in the simulated spatial processes, such as fire spread, seed dispersal, animal movement (Karau and Keane 2007). Finer resolutions provide the greater spatial detail often needed by management (i.e., smaller mapping units and mapscales), but coarser scales reduce simulation times. In an ideal world, spatial and temporal resolutions should match the scale of the ecological processes being modeled and they must also fit with the management objectives. Wildland fire, for example, spreads at fine scales (e.g., < 50 m), whereas climate may vary at much coarser scales (> 1000 m). In the end, however, I've found that most modeling project resolutions are usually determined by available data, computing demands, and the selected model.

Temporal resolutions will also be key in selecting the most appropriate models, but the scale must be consistent with the modeling objective. Several modeling projects in which I participated used models that had annual time steps as the finest temporal resolution, but the modeling objective specifically stated that monthly and daily estimates were needed for reporting. It is important that the finest resolution needed for the modeling objective be identified so that it can be used as a decision threshold in other modeling phases, specifically initialization and parameterization.

How accurate do the results need to be?

There is an old adage that it is impossible to achieve an acceptable balance across good, fast and cheap—people always compromise at least one of the three (Nogami 1982). Most modeling projects have the laudable goal of achieving the highest accuracy possible, but often end up compromising accuracy for time and funding concerns. It is imperative that an accuracy estimate be integrated in the objective or the project specifications, not as a target, but rather as a concept for design. Many decisions made during the modeling project may sacrifice accuracy for expediency so it is important to know when the design of the project achieves an accuracy acceptable for the project. A notion of that threshold should be recognized by the project team. Accuracy, of course, goes hand and hand with the resolution of management decision-making process (previous section).

In summary, it is critical that most modeling projects have a preliminary estimate of the amount of uncertainty that is expected in the modeling results. During parameterization, for example, it is often best to rate the uncertainty in each parameterized value (see Chapter 5). And, when crafting the objective, it's best to think about the amount of uncertainty that is acceptable in results (Table 2.1) and the consequences of making a wrong decision using those model results. Unfortunately, quantifying accuracy is difficult for most models (Chapter 7) because of data limitations, inconsistencies, and incompatibilities. Qualitatively describing uncertainty in the modeling process can go a long way in interpreting simulation results in the proper context.

Selecting the Right Model

The most common question land managers ask ecological modelers is "what model should I use?" This is entirely understandable because the literature is replete with research and management-oriented models that were developed by different people for different purposes and for different ecosystems. It would be impractical to expect land managers to devote enough time to develop a deep knowledge of all available modeling systems—it is just too complicated. Model selection is a complex evaluation of multiple factors that many find difficult and tedious because of their lack of experience in modeling. Therefore, the following is offered as a general guide to help users select the most appropriate model to accomplish the desired task. A more comprehensive selection guide is presented in Keane et al. (2019a). First, a list of questions are presented to define possible criteria for selecting a model (Table 3.1). Please note that many of the criteria for selecting the model trace back to defining the project (previous section).

Table 3.1. The six major constraints that govern the selection of an ecological model for a natural resource management application.

Constraint	Example	What to check?
Appropriateness	Is it able to answer objectives?	Make sure variables match in model and objectives; ensure simulated processes match objective requirements; ensure resolution of the decision matches the resolution of simulated results.
Timeliness	Does it take too long to prepare and run?	High number of variables indicate high levels of parameterization and initialization.
Suitability	Can it be run on existing computer resources?	Inventory available computing resources and match with specifications for the model; interview past users as to requirements; obtain memory and processor speed requirements for model and match to available computers.
Expertise	Are there experts available to help run the model and interpret results?	Are there people that can help running the model or interpreting the results? Is the model easy to learn and understand if there is a lack of experts; are there modelers available to help?
Data	Are there enough data to initialize, parameterize, calibrate and validate the model?	Inventory possible local, regional, or national data sources to determine if available data matches what the model needs for preparation and execution.
Resolution	Are the model's inherent organizational, temporal, and spatial resolutions sufficient?	Match state variable resolution with objectives; determine if a spatial model is needed; match temporal requirements of the objective with model time steps and time spans.

Questions to Define Evaluation Criteria

Which model can answer the modeling objective?

If the modeling objective is well-written, it should be easy to speculate what response variables are needed, and the availability of these variables for output from models should be used as evaluation criteria for model selection. If a model doesn't output those variables or proxies (variables that represent response variables), then it probably shouldn't be selected (**appropriateness** in Table 3.1). For example, if the modeling objective is "to explore changes in tree growth for alternative management scenarios of ...", then the output variables might be diameter, basal area, and tree density by year. This means that only models that output these variables are appropriate.

The model should also simulate those ecological processes important to the project objectives. A prospective model must contain a set of desirable ecosystem components and modules to simulate those processes that are critical to the modeling objective. For example, if the management objective states "evaluate effect of drought on basal area...", then there should be a thorough simulation of the factors that control drought in the model, such as soil water dynamics, vegetation-water interactions, and topological influences. It is suggested that users list all possible ecological processes that they think are critical to their modeling project (see Figures 2.3, 2.7 for example). Then, candidate models can be evaluated against this list to see if they are appropriate for the project.

Which model is most appropriate for project timelines?

All too often, the timeline of the project dictates model selection (**timeliness**). Projects that need to be done quickly will demand simple models that are easier to run. Longer projects can use more complicated models that will widen the scope of the evaluation. Users must remember that the execution time isn't the only time-consuming phase in a modeling project—getting the model ready by implementing the phases detailed in this book takes more time than the actual execution of the model. Unfortunately, it is difficult to assess development and execution times for models that are unfamiliar to the user. The only way to obtain this information is from the literature or in consultation with the model developer, modeling experts, or past model users, so access to modeling expertise is critical for a successful modeling project.

Which model best fits with existing computing resources?

It is crucial that the complexity of an ecosystem model match the available computer resources (**suitability**). A complex model often requires abundant computing resources for normal simulations, which may be inappropriate if only one computer with a single processor is available. Users should identify all possible computers that can used to run the selected ecosystem model and record their specifics and use this as another evaluation criteria.

As mentioned, one of the most important computing resource is the amount of memory available for the model. Many models keep simulated variable values in resident memory instead of writing the values to disk to reduce execution times, so it is important that when this memory fills up, the program doesn't crash or slow to a crawl. Users should address both the memory requirements of available models and the memory resources of the available computers. Sometimes, memory can be up-graded and the cost of upgrading could be also included as evaluation criteria.

A common problem is that some models may need long execution times and may require many computing systems, and this may make it difficult to get the modeling project done on time. Model execution times of over a week are somewhat problematic because there are always small problems that creep into the calibration and execution phases, and long execution times make it difficult to redo simulations if errors are found. Moreover, it is also difficult to keep complex computing systems running for weeks or more due to electrical shortages, operating system bugs, and administration updates. All these factors make it difficult to conduct projects with extensive modeling scenarios (next section). If a complex model is selected, then the user should build extra time into the timeline to anticipate errors in initialization, parameterization, and calibration.

Is there available expertise to run each model?

Look around your agency or organization and determine how many people have the time, knowledge, and experience in complex modeling to help with your project (**expertise**). It may be beneficial to only use a modeling system in which some local people have expertise. This experience criteria can be met by one person, often the user, or by a team of people. A team of people can help in many aspects of the modeling project including providing expertise in parameterization and calibration; development of analysis plans, and review of model results. It is also important that the team have a broad ecological knowledge of the ecosystem(s) being modeled to evaluate the degree of realism needed for model selection. It may be that expert participation needs to be formalized through organizational agreements, contracts, or memorandums of understanding.

Are there enough data to prepare each model?

It would be best to do a quick inventory of available datasets and model input requirements to assess if there are enough data to quantify the initial conditions and parameters of any model (**data**). If data are scarce, then it may not make much sense to select a complex model because high uncertainty in inputs might result in greater uncertainty in outputs (i.e., garbage in, garbage out). Many of the modeling projects that I've been associated with have selected a complex model to answer project objectives only to find that input maps and accurate parameters were mostly unavailable.

Which model best matches the resolution of the objective?

Some modeling objectives can be successfully completed with broad general estimates, and therefore, the resolution of the output variable can be coarse and the model results can be summarized to a generalized value (**resolution**). An example would be that the modeling project only needs the percent of the landscape occupied by a specific community type. Other projects may need complex output for many variables, such plant density, biomass, and productivity estimates summarized by vegetation type and management zones within a simulation landscape. If a project demands these types of detailed reports of ecosystem states, then complex models are often the only alternative.

Next, it is important that the model fulfill the resolution requirements defined in project design. If the model only outputs vegetation cover types when finer resolution variables, such as tree density and basal area, are needed for completion of the objective, then it is time to select another model. If the model outputs values of response variables at an annual time steps when monthly analysis is needed, then strike that model from the list. If the model outputs hundreds of variables when only a "yes" or "no" answer is needed, a simpler and easier model may be more appropriate to save the project a lot of time and money.

It is also important, but not essential, that the detail of simulation be somewhat comparable over all ecological entities being modeled. If the model selected is an LESM, for example, then it is desirable that the same resolution, detail, and accuracy for simulating ecological processes and mapping output variables be consistent wall-to-wall. This means that grasslands should be mapped and simulated at the same resolution as forests, alpine be simulated at the same resolution as temperate forest zones, and rock-lands receive the same treatment as shrublands.

Which model is documented and published?

Users can look to the publication record of a model in the literature to demonstrate its acceptance within the science and management communities. Important items to assess include: (1) has the model been successfully implemented and reported on by others, (2) are model results published in high impact journals, and (3) is the application of the model consistent with the current modeling project?

From a user's perspective, it is important that the model is well-documented, including ample information on using the model and available example input files to demonstrate model set-up. While not critical, a user's manual is priceless for novice users. From a modeler's perspective, it is important that the source code of the model is accessible, well-documented, portable (can be installed on a variety of operating systems), and stand-alone (doesn't require commercial software to run). Open science principles are changing how models are developed and shared, such as Github for the LANDIS model (Tonini et al. 2018) and Frames for wildland fire models (https://www.frames.gov/models). Batch-processing scripts to execute multiple runs of a model are useful for stochastic models that must be run multiple times for a given scenario, and models with input and output structures that are easy to understand facilitate ready parametrization, calibration, and implementation.

How to Evaluate all Models

If possible, the user should first identify a slate of models that might be appropriate for completing the modeling objective. This can come from a review of the literature, but most often, from local modeling experts. Each model can be entered as a row on a spreadsheet. Then the user should identify a list of selection criteria, perhaps by synthesizing criteria from the questions presented above, or by crafting additional criteria that are specific to their project. For example, model flexibility has been important in some of my previous modeling projects, such as the ability of the model to answer questions unrelated to the objective. The language in which the model was programmed; presence of a user-friendly interface to enter input data; ability to link the model input/output to other products; and citation record of the model in the literature are other criteria that may be locally important. Each of these criteria is a column in the spreadsheet. Next, a rating system is created to rate each of the models against each of the criteria—a numbering system with zero being low and 10 being high is a possibility, or a simple low, moderate, and high rating may work in many cases. A weighting system should also be developed to weight each criteria as to its importance to the project objective and logistics. For example, one project might put more weight on scores for existing computing resources than available expertise. And last, the user, or entire modeling team, should rate each of the model using the information at hand. Obviously, model selection is more of an art than a science, but often, if this procedure is followed, it will become obvious that one or two models stand out as good fits to the project.

The most important thing for users to know when selecting a model is that there is no such thing as the perfect model for the project. No model will fit exactly with what all user wants, unless, of course, the model is built specifically for the modeling project. Invariably, compromises must be made and most of modeling projects in which I've participated selected a model that best fit the time, computing, and data constraints rather than the best ecological model available or the most cited model. The limitations of less complicated models were usually accepted so that the project could be successfully finished. Again, there is never a perfect fit for any model to a given modeling project—there are always problems with resolution, data, and output detail.

Delineating the Simulation Landscape

One of the most important aspects in a modeling project is to determine the spatial domain over which to simulate. This spatial domain defines a project's analysis area and will be the spatial context in which to interpret all results. If a stand or point model has been selected, then this section is less important. However, if a LESM is selected, then here are a few recommendations which users might find valuable.

Delineate an area that fits with the modeling objective. It is important that the area to be simulated encompass enough of the right land to meet management objectives while still conforming to common land management mapping structures. For example, if land managers use hydrologic boundaries to define management zones, then the proper simulation area would be one or more watersheds. If the project area consists of a number of polygons, then the simulation area might be a square or rectangle around or within those polygons. The area should be big enough to encompass the spatial footprint of important ecological processes that occur on the landscape while also matching the spatial domain of the management objectives.

Ensure that the area is big enough to describe disturbance processes. A general rule of thumb is that the simulation landscape should be at least five times the size of the largest important disturbance event. For example, if the largest fire in the management area is 100,000 ha, then a landscape of 500,000 ha is needed. This is often difficult to accomplish, given the large size of disturbances in the fire-prone ecosystems of the world and the computing limitations of land management agencies. Often, the available data, computing resources, and time will dictate the size of the simulation landscape, instead of ecological issues. If a smaller area is used, the expression of disturbance regimes on the simulation landscape might be questionable and therefore have greater uncertainty which must be included in the interpretation of model results.

Use round or square landscapes to minimize edge effects. Landscapes that are long and thin have a high perimeter to area ratio and therefore will be subject to extensive edge effects if the selected model explicitly simulates spatial processes (Keane et al. 2003). The edges of the landscape will be missing the influences of those disturbance or vegetation events that are occur outside of the landscape and spread into the landscape. For example, seed dispersal from outside of the landscape perimeter is missing when computing the regeneration potential of stands near the edge of the landscape, or the fire regime is missing those fires that spread into the landscape.

Always use a buffer to minimize edge effects. The most common method to minimize edge effects is to create a *buffer* around the target or context landscape (Figure 3.2). The **target** landscape is the area delineated by management and only output from this area will be summarized to meet the modeling objective. The **simulation** landscape is defined as the entire simulation area, which is the target area plus buffer. As mentioned, the main reason for a buffer is that many spatial processes, such as fire spread and seed dispersal, are often underestimated near landscape boundaries because there is no movement into the landscape from outside the landscape (Figure 3.2) (Keane et al. 2006, Pratt et al. 2006). What happens is that processes from outside the simulation landscape are not spread into the target landscape (immigration) and processes from inside the landscape do not spread outside of the target landscape (emigration). This is important because fires, for example, do not spread into and out of the target landscape, and ignoring this means that the fire regime may not be realistically simulated, especially along landscape edge zones (Figure 3.2). Buffers are a great way to mitigate this problem, even when the simulation landscape is already sufficiently large. It is always best to include buffers on all landscapes when spatially explicit processes are simulated. Most of the landscape simulation projects that I've worked on usually created a buffer 3–5 km wide around target landscapes that were at least 50,000 ha in size. Usually, the bigger the buffer, the more accurate the simulation, but big buffers (> 5 km) usually result in long simulation times and high memory requirements, and as a result, the buffer size is often adjusted to fit a simulation time line. It is important to remember that buffers also need to be initialized and parameterized at the same level of rigor as the target landscape, but all buffer dynamics are ignored when analyzing simulation results to answer the objective.

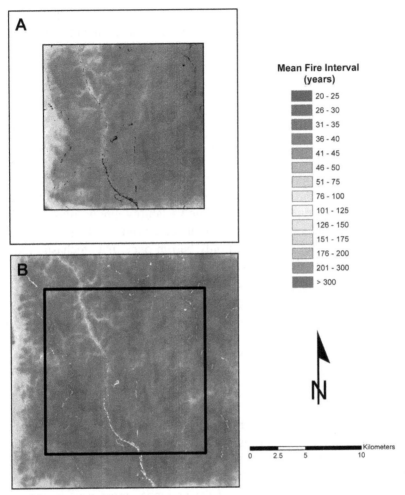

Figure 3.2. Example of edge effects on a square simulation landscape. (A) simulation of a fire regime within the target landscape; (B) the simulation of a fire regime using a 5 km buffer. Note the changes in fire frequency along the buffer-target landscape boundaries. From (Keane et al. 2006).

Creating the Simulation Design

The last stage of project planning is to decide on a simulation design that will fulfill the modeling objectives, and this is often the most critical phase of the design process. Another words, what kind of simulation results should be generated to create the prototype tables and figures that will then be used to fulfill the project objective? Can I just run the model once and be done with the project, or do I need multiple runs in a comprehensive simulation design to fulfill my objectives? In this step a simulation design is created to implement using the selected model to answer the project's objective.

What is important to project design?

The first task is to assess the level of stochasticity (randomness) inherent in the selected model. Highly stochastic models need to be run multiple times (i.e., repetitions=reps) to represent the variability among runs that result from stochastic algorithms. If models are deterministic, then no repetitions are needed and the model need only be run once, but stochastic models absolutely need multiple runs to represent the range of possible responses or predictions. Some landscape models may have minor stochastic modules that seldom influence overall model behavior, and as a result, the relative variability between identical runs may be small (< 5% of the mean) and therefore a lower number of reps are needed. However, most landscape models that simulate disturbance spread over a multitude of years often simulate a great deal of variability across runs (> 25%) because locations of disturbance initiations (e.g., fire start), and the subsequent fire effects, are different across runs (see Figure 2.6 as an example). This range of variability begs the question "how many runs do I need to make to quantify this variation?" If variation is low (< 25% of mean), then perhaps only 3–5 runs are needed, but as the variation increases, there may be a need for at least 10 replications. To be absolutely sure, a confidence interval or sample size power analysis of the variance across runs is needed to estimate the minimum number of runs for a project. Kennedy (2019) has written a wonderful guide on how to determine the number of replicates needed for stochastic models. Sadly, there are few people on the modeling project who know how to do these complicated statistical analyses, and in these cases I would suggest that only five runs are needed if the variation is less than 5%; 10 runs if variation is less than 20%, and 20 runs if variation is > 20%. Often, the number of simulation replications is decided based on project deadlines rather than statistical accuracy or robustness (see later in this section), and it is often much easier to skip the power analysis of variance and just deal with the variation in the statistical analysis and interpretation of model results (Chapter 9).

A second assessment is the reliability of the inputs (initial conditions and parameters) used in the model, especially climate inputs (Table 2.1). When the values of parameters are highly uncertain or unknown, it is suggested that a "scenario" approach be used to bracket the range of uncertainty in the model inputs. A scenario approach is used when the modeler decides how many parameterizations are needed to span the range of uncertainty in model parameters. Let's take climate as an example. Most simulations of future climate impacts need to recognize the wide variation in potential climates as projected by the global circulation models (GCMs). Some GCMs may predict minor ($< 3°C$) increases temperature for a geographic location while other GCMs may predict major increases. Therefore, many landscape and ecosystem modelers often select outputs from 2–5 GCMs to span the range of possible future conditions (Mote et al. 2005, Scholze et al. 2006).

The next assessment concerns the set of management alternatives that are required to fulfill the modeling objectives. Wildfire management scenarios, for example, can include anywhere from zero fires suppressed (historical fire regime) to 98% of the fires suppressed upon initial attack (current target level of management on public lands in US) (Holsinger et al. 2014) (Table 3.2). If management treatments are being modeled in set of a scenarios, such as thinnings, ecosystem restoration cuttings, and timber harvests, then there may be a multiple hierarchy of treatment scenarios (e.g., treatment intensity, area treated, and silvicultural prescription). For example, if the objective is to restore fire to fire-excluded ecosystems, then there may be three treatment designs (silvicultural cuttings, prescribed fire, and a combination of both) that are implemented at three different intensities (no treatment, light treatment, and aggressive treatment where all fire-sensitive species are cut or burned), and at three different levels of extent (zero, 1% of landscape treated per year, and 5% per year). It is important that a "no treatment" or "no action" scenario be included in all simulation designs because landscapes and ecosystems are dynamic and constantly changing even without management actions, and as such, that change must be quantified for land management.

Table 3.2. An example of a comprehensive simulation design for a research project that assessed possible thresholds in fire suppression effort along with the details of the simulation and the list of response variables used in the analysis. Taken from Keane et al. (2019a).

Factor or variable	Levels, parameters, or units	Description
Simulation Design		
Fire Suppression Level	10	Proportion fires suppressed: 0.0, 0.1, 0.2, 0.3, 0.4, 0.5, 0.6, 0.7, 0.8, 0.9 or 10, 20, 30..90% suppression levels.
Fuel Treatment Effort	4	Mechanical thinning of fire sensitive species and subsequent prescribed burns at the following levels of implementation: none (N), 3% landscape treated (Business as Usual-BAU), 10% landscape treated (fully funded fuel treatment program FF); treat all lands that need it (no holds barred NHB).
Climate	2	Historical climate (HIST) and the hot, dry RCP8.5 IPCC (2011) scenario for the US northern Rocky Mountains (RCP8) where temperatures are around 4.5°C warmer and precipitation is around 90% of historical climate: Offsets were T_{max} 4.3, 5.5, and 6.8°C and T_{min} 4.8, 6.8, 6.8°C for EFBR, YCP, CROWN, respectively, and precip 200%, 110%, and 120% more.
Landscape	3	East Fork Bitterroot River (EFBR), Yellowstone Central Plateau (YCP), Crown of Continent (CROWN) (Figure 2).
Simulation Specifications		
Simulation length	200 years	Simulate landscape dynamics for 200 years to get a full representation of the fire and fire management regimes.
Simulation Replications	10 reps	Needed to get sufficient observations to detect thresholds but low enough not to generate too many observations for statistics.
Simulation output interval	10 years	Generated stand, site, and landscape level output at decade intervals; used the last 100 years (10 observations) in the analysis.
Time step	Daily	Cycle the 50+ year weather records for each landscape for 200 years.

Table 3.2 contd. ...

... Table 3.2 contd.

Factor or variable	Levels, parameters, or units	Description
Response variables		
BA	(m^2 ha^{-1})	Basal area of all trees across the landscape.
CBD	(kg m^{-3})	Canopy bulk density.
CWD	(kg m^{-2})	Coarse woody debris-downed woody material greater than 10 cm diameter.
FWD	(kg m^{-2})	Fine woody debris-downed woody material less than 10 cm diameter.
TC	(kg m^{-2})	Total carbon of living and dead biomass.
PLFAS	(%)	Percent of the landscape composed of fire adapted species.
PLBURN	(%)	Percent of the landscape burned annually.

How do I design the simulation project?

Most simulation designs use scenario-based approaches that are analyzed for significant differences using statistical techniques (e.g., ANOVA, MANOVA). A multiple factorial design is often used as a simulation project design where all factors (e.g., climate, management actions) are evaluated for significant differences across all the levels (e.g., intensities, area treated, treatment designs). Each combination of factor and level is termed a **scenario** in this book. In our examples above, you may want a simulation design where two management treatments alternatives (no treatments and a prescribed fire on 1% of the landscape per year) is evaluated under two wildfire management scenarios (zero suppression and full suppression) and three climate scenarios (historical, RCP4.5, RCP8.5) (Table 3.2). Simulation designs can become quite complex with multiple hierarchical scenarios and many levels resulting in numerous scenarios and many model executions. For example, four management scenarios (no action, timber harvest only, restoration treatment only, and harvest+restoration) under three fire suppression levels (no suppression, 50% suppression, and 98% suppression) and two exotic scenarios (with and with exotic species) by three climates (see above) with 10 replications would require 720 model executions (4x3x2x3x10=720).

It may be useful to provide some examples of various simulation designs from modeling projects published in the literature along with the detailed example in Table 3.2. In one model comparison projects, Cary et al. (2006) evaluated area burned simulated from five models using a simulation design that included ten replicates of three factors—topography (two levels—flat and dissected), weather (constant weather and highly variable), and climate (historical and hot, dry)—to determine the sensitively of these factors to fire dynamics. Several research modeling projects used three main factors—climate (historical, warm-dry, hot-dry), fire management (no suppression, 50% fires suppressed, and 98% suppressed, and fuel treatment (no treatment, 3% per year treated, 10% per year treated)—to assess changes in fire dynamics, vegetation distribution, and landscape pattern and structure (Loehman et al. 2011, Holsinger et al. 2014, Clark et al. 2017).

It will be incredibly tantalizing for many readers to include numerous factors with many levels in the simulation design to really give punch to a modeling project. However, users should give some focused thought as to: (1) how long it will take to perform these simulations, (2) how much output will the runs produce; and (3) how many people have the expertise to perform the subsequent analysis for this design? Remember, the number of runs needed to complete a modeling project increases by a product rather than a sum. Using the design in a previous paragraph as an example, the number of runs to complete the project would be 720 simulations, and, if each run takes a day to complete on a laptop computer with only one processor, it would take too long to complete the project. As mentioned previously, it is usually available computer resources, not the statistical requirements that dictate simulation design. A parsimonious simulation design is always best as long as it answers the simulation objective.

There are a few things you can do to make the simulation design more effective. First, it is important that the range of levels in any one factor encompass and capture the scope of decision or objective, especially if this is a research project. An example of this can be found in the Keane et al. (2019b) study where four levels of fuel treatment were simulated to find the tipping point of how much land must be treated to manage wildland fire. They simulated 0, 1, 5, and 100 percent of the landscape treated and found that the tipping point was somewhere between

5 and 100 percent. They probably could have better estimated the tipping point had they picked better levels or simulated more scenarios. The most common mistake people make is designing alternative management scenarios that produce results that are not statistically significant. Another suggestion is to design a simulation experiment so that it has a base case against which all other scenario results can be compared. In most management designs, the base case is often the no action scenario, but sometimes the base case is a scenario that captures the historical range and variation (Keane 2012), or it may be the "preferred" management scenario. And last, of course, it is important that the scenarios capture those elements that are important to the simulation objective.

Implementing a Successful Project

There are many tasks that will maximize the success of your project. First, it is strongly advised that each phase of the simulation project be extensively documented so that the analysis and report writing phases are straightforward (Chapter 9). Graphics and word processing software, for example, can be used to document the findings of each of the phases of the project. A notebook, blog, or audio recording could also be used to document each phase. The most important items to document are any changes to the: (1) project's objective, (2) agreed upon procedures, or (3) results summaries. Also important are any limitations or strengths of the model that are encountered during the modeling project. For example, the execution times of the model should be recorded as reference in the execution phase; the sources of all parameters should be recorded to document the parameterization; and the directory structure of both input and output files (Chapter 8) should be specified for ease of analysis (Chapter 9).

Next, it is advised that a team of people be identified to help with all model phases. If available, this includes the developer of the model; past users of the model; other resource professionals; and ecologists. These people can be consulted during the preparation (initialization, parameterization, calibration) and analysis phases. It is also important that there are people who understand the ecosystem and provide the expertise to evaluate realism in model predictions, especially during the calibration phase.

Collaborative, iterative modeling approaches, such as that detailed by Gustafson et al. (2006), have been quite successful in many modeling projects; these approaches depends on collaboration among model experts, resource experts and decision-makers.

In summary, the design of a successful modeling project includes crafting a succinct set of objectives; defining the bounds of the project; selecting the best model; selecting the simulation landscape; and then crafting the simulation design in that order. Hopefully, these steps will make project planning less stressful and more informative. Again, a well-designed modeling project defines the boundaries and context for all future decisions that will invariably need to be made in the six phases of a modeling project.

References

Bovend'Eerdt, T. J. H., R. E. Botell and D. T. Wade. 2009. Writing SMART rehabilitation goals and achieving goal attainment scaling: A practical guide. Clinical Rehabilitation 23: 352–361.

Cary, G. J., R. E. Keane, R. H. Gardner, S. Lavorel, M. D. Flannigan, I. D. Davies, C. Li, J. M. Lenihan, T. S. Rupp and F. Mouillot. 2006. Comparison of the sensitivity of landscape-fire-succession models to variation in terrain, fuel pattern, climate and weather. Landscape Ecology 21: 121–137.

Clark, J. A., R. A. Loehman and R. E. Keane. 2017. Climate changes and wildfire alter vegetation of Yellowstone National Park, but forest cover persists. Ecosphere 8: e01636-n/a.

Gregory, R., L. Failing, M. Harstone, G. Long, T. McDaniels and D. Ohlson. 2012. Structured decision making: a practical guide to environmental management choices. John Wiley & Sons.

Gustafson, E. J., B. R. Sturtevant and A. Fall. 2006. A collaborative, iterative approach to transferring modeling technology to land managers. pp. 43–64. *In*: A. H. Perera, L. Buse and T. R. Crow (eds.). Forest Landscape Ecology: Transferring knowledge to Practice. Cambridge Press, London, UK.

Hann, W. J., J.L. Jones, M.G. Karl, P. F. Hessburg, Robert E. Keane, D.G. Long, J.P. Menakis, C. H. McNicoll, S. G. Leonard, R. A. Gravenmier and B. G. Smith. 1997. An assessment of ecosystem components in the Interior Columbia Basin and portions of the Klamath and Great Basins Volume II—Landscape dynamics of the Basin. General Technical Report PNW-GTR-405, USDA Forest Service Pacific Northwest Research Station.

Holsinger, L., R. E. Keane, D. J. Isaak, L. Eby and M. K. Young. 2014. Relative effects of climate change and wildfires on stream temperatures: A simulation modeling approach in a Rocky Mountain watershed. Climatic Change 124: 191–206.

Karau, E. C. and R. E. Keane. 2007. Determining landscape extent for succession and disturbance simluation modeling. Landscape Ecology 22: 993–1006.

Keane, R. E., D. G. Long, J. P. Menakis, W. J. Hann and C. D. Bevins. 1996. Simulating coarse-scale vegetation dynamics using the Columbia River Basin succession model:

CRBSUM. RMRS-GTR-340, U.S. Dept. of Agriculture Forest Service Intermountain Research Station, Ogden, UT.

Keane, R. E., G. J. Cary and R. Parsons. 2003. Using simulation to map fire regimes: An evaluation of approaches, strategies, and limitations. International Journal of Wildland Fire 12: 309–322.

Keane, R. E., L. Holsinger and S. Pratt. 2006. Simulating historical landscape dynamics using the landscape fire succession model LANDSUM version 4.0. General Technical Report RMRS-GTR-171CD, USDA Forest Service Rocky Mountain Research Station, Fort Collins, CO USA.

Keane, R. E. 2012. Creating historical range of variation (HRV) time series using landscape modeling: overview and issues. pp. 113–128. *In*: J. A. Wiens, G. D. Hayward, H. S. Stafford and C. Giffen (eds.). Historical Environmental Variation in Conservation and Natural Resource Management. John Wiley and Sons, Hoboken, New Jersey.

Keane R. E., R. A. Loehman and L. M. Holsinger. 2019a. Selecting a Landscape Model for Natural Resource Management Applications Current Landscape Ecology Reports doi:10.1007/s40823-019-00036-6.

Keane R. E., K. Gray, B. Davis, L. Holsinger and R. A. Loehman. 2019b. Evaluating ecological resilence across wildfire suppression levels under climate and fuel treatment scenarios using landscape simulation modeling. International Journal of Wildland Fire 34: 1–15.

Keeney, R. L. 2007. Developing objectives and attributes. Advances in decision analysis: From foundations to applications: 104–128.

Kennedy, M. C. 2019. Experimental design principles to choose the number of Monte Carlo replicates for stochastic ecological models. Ecological Modelling 394: 11–17.

Loehman, R. A., J. A. Clark and R. E. Keane. 2011. Modeling Effects of Climate Change and Fire Management on Western White Pine (Pinus monticola) in the Northern Rocky Mountains, USA. Forests 2: 832–860.

Lutes, D. C., R. E. Keane, J. F. Caratti, C. H. Key, N. C. Benson, S. Sutherland and L. J. Gangi. 2006. FIREMON: Fire effects monitoring and inventory system. General Technical Report RMRS-GTR-164-CD, USDA Forest Service Rocky Mountain Research Station, Fort Collins, CO USA.

Mote, P. W., E. P. Salath and C. Peacock. 2005. Scenarios of future climate for the Pacific Northwest. A report prepared for King County (Washington)'s October 27, 2005 climate change conference. The Future Ain't What It Used to Be: Preparing for Climate Disruption. Climate Impacts Group, Center for Science in the Earth System, Joint Institute for the Study of the Atmosphere and Ocean, University of Washington, Seattle USA.

Nogami, G. Y. 1982. Good-fast-cheap: Pick any two: Dilemmas about the value of applicable research 1. Journal of Applied Social Psychology 12: 343–348.

Pratt, S. D., L. Holsinger and R. E. Keane. 2006. Modeling historical reference conditions for vegetation and fire regimes using simulation modeling. General Technical Report RMRS-GTR-175, USDA Forest Service Rocky Mountain Research Station, Fort Collins, CO USA.

Scholze, M., W. Knorr, N. W. Arnell and I. C. Prentice. 2006. A climate-change risk analysis for world ecosystems. Proceedings of the National Academy of Sciences 103: 13116–13120.

Tonini, F., C. Jones, B. R. Miranda, R. C. Cobb, B. R. Sturtevant and R. K. Meentemeyer. 2018. Modeling epidemiological disturbances in LANDIS-II. Ecography 41: 2038–2044.

4

Initialization
How to Begin a Simulation

"Whenever I start a project, I have a broad range of possibilities."

Shaun Tan

Initialization—the process of assigning starting values to modeled variables.

ABSTRACT

Initialization is the task of estimating and assigning initial values to all the state variables in the simulation model used in the project. This chapter covers the quantification of the initial conditions of the landscape or ecosystem to begin a simulation. Three types of initialization are covered—forward, past, and representative initializations. Then the six steps of initialization are provided including identifying input variables, collecting data for initialization of those variables, formatting the data for input to the model, running the model with newly created initialization, evaluating the results for possible mistakes, and revising any initial values that are in error. Last, a comprehensive set of problems that might arise during initialization are presented and solutions are suggested.

Introduction

Initialization is the task of estimating and assigning initial values to all state variables for the simulation model used in the project. It involves searching for and obtaining data that describes the conditions of the plants, stands, communities, and regions in the simulation landscape, and compiling that data in the appropriate format, to start and execute the model. Models rarely require the user to input every initial value; usually models utilize a subset of user-entered information to initialize the remaining values in the model. For example, a landscape model might use a list of tree diameters, heights, and species to compute root, stem, and foliage carbon pools based on allometric equations found in the literature. Initialization does not include the quantification of model parameters; that is the subject of the next chapter. In this chapter, we deal with the quantification of the initial conditions of the landscape or ecosystem to begin a simulation. All initializations depend on the project objectives and simulation design (Chapter 3); this chapter assumes that the reader has designed a modeling project and identified an appropriate model for use.

Some feel that initialization is the least important of all the modeling phases, but often, they would be sadly mistaken—initialization is every bit as important as all the other phases in this book for specific modeling projects. This is because there are some major consequences of initialization for evaluation of simulation output. A common dilemma I've often encountered when initializing landscapes in fire-prone ecosystems, for example, is that today's landscapes often represent the influence of a century of fire exclusion; contemporary forest conditions, such as tree densities, fuel loadings, and standing biomass, are a result of decades of fire suppression. Therefore, if the modeling objective is to simulate the historical fire regime and the user employed today's conditions as initial values, the effects of fire exclusion would be evident in the output for at least a century of simulation (Keane et al. 2006). Output from the first years of simulation may appear to be in error because there will be major declines in ecosystem biomass as fires burn on the simulated landscape using historical fire return intervals. This dilemma has sometimes caused modelers to start a simulation from "bare ground" by assigning a zero values to many of the dynamic state variables then running the model until it reaches an "equilibrium" to populate the initial values. This may be a good idea in landscapes or ecosystems with minor disturbances or if the model is indeed an equilibrium model, but, in most ecosystems of the world, disturbances and changing climates complicate dynamic equilibriums, especially for complex mechanistic models, making bare ground initializations inappropriate or unrealistic.

Another example of the importance of initialization is found in a project that simulated restoration in whitebark pine ecosystems of North America (Keane et al. 2017). The exotic disease white pine blister rust killed many whitebark pine trees over the last two or three decades in many portions of the tree species' range (Kendall and Keane 2001). Therefore, many trees sampled in the field to obtain initialization data were dead from the disease. As a result, it was difficult to actually determine the historical dynamics of whitebark pine populations when these snags were input into a model as initial conditions. Loehman et al. (2011) modified the snag data to make them "living" trees in their simulations to better represent historical landscape dynamics and to reduce simulation time.

Initialization is also important for the calibration (Chapter 6) and validation (Chapter 7) modeling phases. Often, an evaluation of model output for realism requires that the model results be compared with observed reference conditions. The problem, of course, is that the data used to initialize the model represent today's conditions and not the historical conditions that led to the formation of the reference data. For example, let's assume that the only data set for model evaluation is a 20 year record of streamflow data that spans the years 1990–2010. Stand and tree inventory records for the year 1990 are often missing or incomplete so it is common that initialization data represent conditions at or near 2010 as used as a compromise. However, when if the user employed the 2010 conditions to represent the 1990 initial conditions, the resultant validation time series would have some serious limitations in that major landscape changes took place during those 20 years as a result of tree growth, disturbance, and human land use. In these cases, modelers often "grow" the landscape backward using reformulated allometric equations and/or trial-and-error iterative techniques to approximate the 1990s conditions (see Chapter 7). Indeed, initialization is an important task in any modeling project and the user needs to know how best to initialize the model for their modeling project.

Initializing the Model

In general, three types of initialization are used for most ecological simulations. First, the model uses today's conditions and simulates forward in time. Second, the model uses data from the past, either historical data or manipulated data, and simulates to the present, and perhaps, into the future. As mentioned, this is primarily used as an initialization for validation (Chapter 8). And last, the model is initialized with "representative" data that do not reflect current or past conditions, but instead, reflect a standardized starting place for all simulations.

Representative initializations are quantified using a wide variety of techniques. One technique is to use manipulated or mocked-up data to develop a series of initializations to capture a wide range of starting conditions to bracket some modeling objective. Xiao et al. (2000), for example, developed several "bogus" initializations to understand hurricane dynamics. Another technique is the "spin up" of an equilibrium model and use values from the last year of spin-up simulation as initialization values (Thornton and Rosenbloom 2005). Still another approach is to run a model for a period of time using contemporary or past initializations and then using the values from that landscape at a pre-determined time as an initialization for the project's simulations. The FireBGCv2 model, as an example, allows the user to save results from any simulation year for an initialization (Keane et al. 2011). Landscape modelers often use "neutral" landscapes (contrived data layers) as initial conditions in landscape simulation experiments (Gardner et al. 1987). Cary et al. (2006), for example, used a checkerboard pattern to represent vegetation community ages on a neutral landscape to understand fuel treatment effects on fire regimes. At any rate, most initializations in natural resource management projects involve a representation of contemporary conditions.

The main challenge in the initialization process is to decide what initial methods and conditions to use. Knowing the type of model helps define the initialization process. As mentioned, equilibrium models may be initialized by zeroing out all state variables and then running the model until results obtain equilibrium (Thornton and Rosenbloom 2005). Other models must have realistic numbers to start the state variables otherwise chaotic results may result. Initialization will also depend on the modeling objective. If simulations are meant to characterize future conditions, then it is essential that the model be initialized using data that represent today's conditions. However, if the model is used in a comparison exercise across various scenarios, then it may be acceptable to assign initial values to state variables based on the range of possible outcomes for each variable. For example, one may want to set the initial value for a variable based on a mean of past simulations or field data. Once the model has been selected and the modeling objective stated, then the following initialization steps should be somewhat easy.

The steps involved in model initialization are similar across all models regardless of model complexity. In this book there are basically seven steps in model initialization (Figure 4.1). These steps don't need to be done in order, but all of them need to be completed to properly initialize a model.

Identify input variables

The state variables that need to be initialized must first be identified. This can be done by referencing the user's manual for the selected model and following the modeler's instructions. Unfortunately, many stand and landscape models were developed for research applications and lack detailed user's guides or instructional material. In these cases, the best approach to identifying variables is to obtain previously created input files that were used in other projects for that model. It is often difficult to determine the type of variable and its units from these files, especially if they were poorly documented. If old input files are missing or difficult to decipher, then the only alternative is to dive into the computer code and figure it out from the structure of the input routines. Hopefully, the modeler has extensively documented the code so understanding the code will be easy. Ideally, the modeler should be around for consultation and

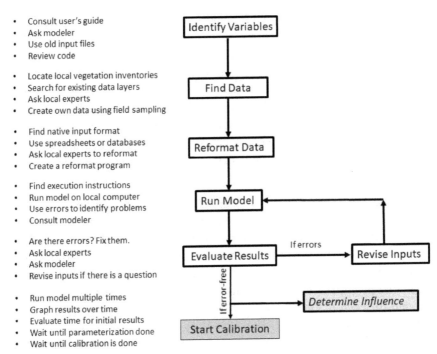

- Consult user's guide
- Ask modeler
- Use old input files
- Review code

- Locate local vegetation inventories
- Search for existing data layers
- Ask local experts
- Create own data using field sampling

- Find native input format
- Use spreadsheets or databases
- Ask local experts to reformat
- Create a reformat program

- Find execution instructions
- Run model on local computer
- Use errors to identify problems
- Consult modeler

- Are there errors? Fix them.
- Ask local experts
- Ask modeler
- Revise inputs if there is a question

- Run model multiple times
- Graph results over time
- Evaluate time for initial results
- Wait until parameterization done
- Wait until calibration is done

Figure 4.1. The steps that are involved in the initialization of ecosystem models. Steps in italics indicate that the step is optional. Various tasks are summarized to the left of each step.

help during this process. Many people prioritize the availability of the modeler and detailed documentation as an important criteria in model selection.

Collect data

Once the input state variables have been identified, data must be collected to quantify initial conditions for those state variables. This is often a difficult task because it is rare that there are sufficient data to initialize a model for a management application. It is rarer still to find data that match the scale and resolution of complex ecosystem models. Spatial models, such as LESMs for example, demand extensive input data layers (GIS maps) of a specific resolution and extent, and these layers, whether raster or vector, must represent landscape composition and structure for the start of the simulation. Plot data is usually needed to assign starting values to simulated composition and structure attributes, and digital maps need to be cross-walked so that the mapped values reference each data layer categories. For example, field measured plant biomass estimates collected on a plot with a geo-referenced location can be used to assign biomass estimates to a classification category represented in an input data layer, such as cover type.

Most modelers have found that a generalized scheme must be developed to guide the initialization process when there are inconsistent and spotty data. Such schemes should be designed to match the quality and coverage of the data to the scale and type of input variables needed for initialization. This is best illustrated by an example. To initialize the FireBGCv2 model for modeling projects, a tree list (number and sizes of trees per unit area) was needed for each mapped stand on the simulation landscape (Keane et al. 2011). When few existing timber inventory databases were found, Keane et al. (2011) created two spatial data layers—biophysical setting and cover type—to guide the assignment of tree lists to all landscape polygons and pixels (Holsinger et al. 2014, Clark et al. 2017). The biophysical settings layer mapped a potential vegetation type classification where categories represent unique biophysical conditions (Pfister and Arno 1980), and the cover type layer was taken from existing map layers describing vegetation composition based on basal area (Eyre 1980). Then, Forest Inventory and Analysis (FIA) inventory field plots on or near that landscape were keyed to the mapped potential vegetation type and cover type combinations and FIA tree data recorded for a modal plot was assigned to each stand (Loehman et al. 2017). The search area was expanded to search for FIA plots for those habitat type-cover type combinations that had no plots. Many variations of this approach are possible, such as using timber inventory data instead of FIA plots, or adding in a structural stage layer to more accurately map the vegetation mosaic, or using topographic settings (combinations of elevation, aspect, and slope classes) to map biophysical variation instead of potential vegetation types.

There are many sources for potential data to initialize models, but they vary in availability, quality, and quantity by geographic area, time period, and land ownership, so a comprehensive guide for selecting initialization data sources is impossible. However, there are a few data sources for initializing landscape models in the US that warrant mention. The LANDFIRE spatial data layers (Keane et al. 2007, Rollins 2009) maintained by EROS Data Center (https://www.landfire.gov/) has some excellent layers to use to represent biophysical conditions, vegetation composition, vegetation structure, fuel loadings, fire regimes, and many other interesting attributes that can be used for initialization. As mentioned, the FIA program provides extensive plot data collected on 5 km by 5 km grid throughout the forested US; these data have both vegetation and fuels information that can be used for a wide variety of initialization tasks (Jenkins and Birdsey 1998, Shaw et al. 2005, Woodall and Williams 2005) (https://www.fia.fs.fed.us/tools-data/index.php).

If there is ample time and funding to conduct a field campaign for collecting initialization data, then here are some important suggestions that might make that effort more effective:

1. Confirm that the sampled data can be used for initialization. Too often, a sampling effort was implemented for a modeling project and the collected data was incompatible for initialization because of scale inconsistencies, improper units, and sampling compromises. One common problem in our modeling projects was that tree regeneration was ignored or inappropriately sampled for model initialization.

2. Collect ancillary data that may be used to assist in initialization. Elevation, aspect, slope, or land use could be assessed in the field and then used to stratify results to aid in initializing the model from scanty data.

3. Merge data collection efforts to help with three other modeling phases (parameterization, calibration, and validation). Since the bulk of sampling costs are in travel, it is important that data useful to all phases be collected during one field effort (see Chapters 5, 6, 7).

4. Ensure collected field data are also useful to your organization. Collect ancillary data that may be helpful to your organization, especially data that can complement the management actions that

are simulated in the modeling project. Wildland fuel data, for example, can be collected along with tree population data if the preferred management action that resulted from the modeling project suggested that prescribed burning is the best management alternative —the collected fuels data can be used to create the fire prescriptions (i.e., desired conditions for burning).

5. Use standardized methods and protocols that mesh with organizational requirements. It is highly desirable to merge the collected data into corporate databases once the initialization is finished. These data may be especially useful in other organizational projects.

If there are no corporate data standards, a number of standardized methodologies could be used as templates for field sampling in validation field campaigns. The FIREMON (Lutes et al. 2006) and FFI (Lutes et al. 2009) packages are tools that could be used in field sampling campaigns. The selected system should include sampling designs, protocols, field codes, databases, and possible analysis reports.

Format data

Once the data are collected, they must be reformatted to fit the input structure of the model. This is often one of the most odious tasks in initialization because if data are improperly reformatted, the next step (running the model) can be quite challenging. If the reader is skilled in programming, then the data can be reformatted using specially designed programs written in various computer languages or scripts, such as C++, Python, or Fortran. If this isn't possible, then data can be imported into other data management packages, such as relational databases or spreadsheets, and the reformatting can be done using routines native to that software. Many modeling projects have reformatted the collected data into ASCII files using a standard text editor, spreadsheet, or word processor.

There are three aspects of formatting the data that are important. The first is the most obvious—make sure the data are in the model's required format. The second is less obvious but just as important—make sure the data are in the right units. Too often, the model will require metric units while inventory data were in English units. And last, it is important to make sure the data are understandable to the user and the modeling team; the initialization data should be thoroughly documented and comprehensive meta-data information should be developed so that changes to the initialization dataset can be easily done.

Run model

The simplest step in this phase is to execute the model to see if the input values are suitable for the model. The user simply runs the model using the recently created reformatted data files as input. The biggest challenge in this step is that when the model will not run (i.e., crash), it may be difficult to find the input data value that is in error. It could be that the values are in the wrong columns or the values contain special characters that the model can't recognize. If the modeler included extensive error statements in the program, then finding all input errors may be somewhat easy. However, many research models have incredibly poor error statement structures and the model simply crashes without a notice or reason. In this case, a trial-and-error method is used to identify erroneous values.

Evaluate results

Now comes the critical process of determining if the input data are acceptable for future modeling tasks. This is different from running the model because it involves detecting if the model correctly accepted the initial conditions and there are no apparent errors in the model execution. Model output for the first or more years should be carefully assessed to determine if there are any bizarre results that might indicate an erroneous initial value.

Revise inputs

If the evaluation described in the previous step finds that there are indeed problems with the input files, then the data need to be *revised* and the process of running and evaluating the model need to be repeated. These steps are repeated until the user has determined that the initialization is acceptable (no obvious errors) and other phases can be started.

Initialization Concerns

Often, novice modelers will spend an inordinate amount of time collecting, synthesizing, and preparing data for model initialization because they want the data to be the best possible representation of initial conditions. While this is commendable, it probably isn't as important as some of the other modeling phases or some other facets of initialization. In fact, initialization is one of the modeling phases in which the timelines can be modified to fit project timelines. One reason for this is that initialization data defines a starting point for modeling and the data are only important for the first few decades of simulation. Minor uncertainties in initialization make little difference toward the end of a 500 year simulation (Keane et al. 2011). However, studies have shown that large errors in initialization may influence a major slice of the initial simulation time span (Keane et al. 2002, Keane et al. 2006). Therefore, initialization, similar to all the other phases of modeling, is more an art than a science when timelines are tight. Small compromises during each initialization step (Figure 4.1) must be made to match the available data with the level of resolution needed to minimize the influence of initial conditions on model results.

If the model selected is an LESM, then it is important that the initialization use the same resolution, detail, and accuracy for mapping initial conditions across the simulation landscape—wall-to-wall. This means that grasslands should be mapped at the same resolution as forests, alpine be mapped at the same resolution as temperate zones, and rock-lands receive the same treatment as shrublands.

Determine Influence of Initial Conditions

Once the initialization has been deemed acceptable and the user feels that other modeling tasks can begin, an additional step is suggested but not required. The user should run the model multiple times and graph the results over time to determine the influence of initial conditions on the beginning years of the simulation, and if there is influence, users need to determine when the influence of initial conditions is minimal over the simulation time span. This step should be done once the parameterization

(Chapter 5) and calibration (Chapter 6) tasks are finished. This step is done to estimate the impact of initial conditions on subsequent results.

As mentioned above, inappropriate initial conditions may lead to questionable simulation results. If the initial conditions accurately portray

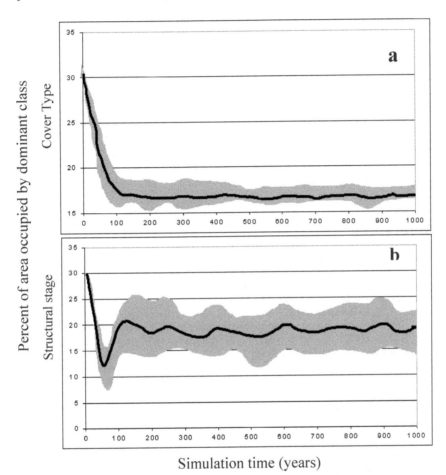

Simulation time (years)

Figure 4.2. Effects of initial values on model output from the Keane et al. (2002) LANDSUM model initialization experiment. Note that the initial conditions at year 1 are quite different than when the model comes into equilibrium. In this case, the model LANDSUMv4 was being used for the approximation of historical conditions and the starting conditions were taken from contemporary landscape that had over 100 years of fire exclusion. It appears the first 100 years need to be removed from the simulated time series to properly describe historical dynamics.

91

the state of the ecosystem or landscape for the start of a simulation, then no further work needs to be done to mitigate the effects of initial conditions. However, if initial conditions are inconsistent with simulation goals, such as in the previous fire exclusion example, then effects of initial conditions must be mitigated. Keane et al. (2006), for example, found that poor initializations affected up to 100 years in LANDSUM simulations (Figure 4.2).

Perhaps the only way to mitigate adverse effects of inappropriate initial conditions when simulating historical regimes or past landscape dynamics is to eliminate the first set of years from the analysis. A trial and error method is often used to determine how many years to remove from the simulated results. To do this, the user would run the model for hundreds of years and annual output would be graphed over time (Figure 4.2). The user then visually inspects the output to determine when the effects of initial conditions are the least evident.

References

Cary, G. J., R. E. Keane, R. H. Gardner, S. Lavorel, M. D. Flannigan, I. D. Davies, C. Li, J. M. Lenihan, T. S. Rupp and F. Mouillot. 2006. Comparison of the sensitivity of landscape-fire-succession models to variation in terrain, fuel pattern, climate and weather. Landscape Ecology 21: 121–137.

Clark, J. A., R. A. Loehman and R. E. Keane. 2017. Climate changes and wildfire alter vegetation of Yellowstone National Park, but forest cover persists. Ecosphere 8: e01636-n/a.

Eyre, F. H. E. 1980. Forest cover types of the United States and Canada. Society of American Foresters, Washington DC., USA.

Gardner, R. H., B. T. Milne, M. G. Turner and R. V. O'Neill. 1987. Neutral models for the analysis of broad-scale landscape pattern. Landscape Ecology 1: 19–28.

Holsinger, L., R. E. Keane, D. J. Isaak, L. Eby and M. K. Young. 2014. Relative effects of climate change and wildfires on stream temperatures: a simulation modeling approach in a Rocky Mountain watershed. Climatic Change 124: 191–206.

Jenkins, J. C. and R. Birdsey. 1998. Validation databases for simulation models: above ground biomass and net primary productivity (NPP) estimation using easwide FIA data. General Technical Report NC-212, USDA Forest Service, North Central Research Station, Boise, ID., USA.

Keane, R. E., L. Holsinger and S. Pratt. 2006. Simulating historical landscape dynamics using the landscape fire succession model LANDSUM version 4.0. General Technical Report RMRS-GTR-171CD, USDA Forest Service Rocky Mountain Research Station, Fort Collins, CO USA.

Keane, R. E., L. M. Holsinger, M. F. Mahalovich and D. F. Tomback. 2017. Evaluating future success of whitebark pine ecosystem restoration under climate change using simulation modeling. Restoration Ecology 25: 220–233.

Keane, R. E., R. A. Loehman and L. M. Holsinger. 2011. The FireBGCv2 landscape fire and succession model: a research simulation platform for exploring fire and vegetation dynamics. General Technical Report RMRS-GTR-255, U.S. Department of Agriculture, Forest Service, Rocky Mountain Research Station, Fort Collins, CO USA.

Keane, R. E., R. Parsons and P. Hessburg. 2002. Estimating historical range and variation of landscape patch dynamics: Limitations of the simulation approach. Ecological Modelling 151: 29–49.

Keane, R. E., M. G. Rollins and Z. Zhu. 2007. Using simulated historical time series to prioritize fuel treatments on landscapes across the United States: the LANDFIRE prototype project. Ecological Modelling 204: 485–502.

Kendall, K. C. and R. E. Keane. 2001. Whitebark pine decline: Infection, mortality, and population trends. pp. 221–242. Whitebark Pine Communities: Ecology and Restoration. Washington D.C.: Island Press c2001.

Loehman, R. A., A. Corrow and R. E. Keane. 2011. Modeling climate changes and wildfire Interactions: Effects on whitebark Pine (Pinus albicaulis) and implications for restoration, Glacier National Park, Montana, USA. pp. 176–188. *In*: The Future of High-Elevation, Five-Needle White Pines in Western North America: Proceedings of the High Five Symposium. U.S. Department of Agriculture, Forest Service, Rocky Mountain Research Station, Missoula, MT.

Loehman, R. A., R. E. Keane, L. M. Holsinger and Z. Wu. 2017. Interactions of landscape disturbances and climate change dictate ecological pattern and process: Spatial modeling of wildfire, insect, and disease dynamics under future climates. Landscape Ecology 32: 1447–1459.

Lutes, D. C., N. C. Benson, M. Keifer, J. F. Caratti and S. A. Streetman. 2009. FFI: A software tool for ecological monitoring*. International Journal of Wildland Fire 18: 310–314.

Lutes, D. C., R. E. Keane, J. F. Caratti, C. H. Key, N. C. Benson, S. Sutherland and L. J. Gangi. 2006. FIREMON: Fire effects monitoring and inventory system. General Technical Report RMRS-GTR-164-CD, USDA Forest Service Rocky Mountain Research Station, Fort Collins, CO USA.

Pfister, R. D. and S. F. Arno. 1980. Classifying forest habitat types based on potential climax vegetation. Forest Science 26: 52–70.

Rollins, M. G. 2009. LANDFIRE: A nationally consistent vegetation, wildland fire, and fuel assessment. International Journal of Wildland Fire 18: 235–249.

Shaw, J. D., B. E. Steed and L. T. DeBlander. 2005. Forest Inventory and Analysis (FIA) annual inventory answers the question: What is happening to pinyon-juniper woodlands? J. For 103.

Thornton, P. E. and N. A. Rosenbloom. 2005. Ecosystem model spin-up: Estimating steady state conditions in a coupled terrestrial carbon and nitrogen cycle model. Ecological Modelling 189: 25–48.

Woodall, C. and M. Williams. 2005. Sampling protocol, estimation, and analysis procedures for the down woody materials indicator of the FIA program. General Technical Report NC-256, USDA Forest Service, North Central Research Station, St. Paul, MN, USA.

Xiao, Q., X. Zou and B. Wang. 2000. Initialization and simulation of a landfalling hurricane using a variational bogus data assimilation scheme. Monthly Weather Review 128: 2252–2269.

5

Parameterization

How to Tune the Model for Local Applications

"However useful computer models may be, the one thing they cannot be is evidence. Computer climate models are simply conjectures."

Nigel Lawson

Parameter—a numerical or other measurable factor forming one of a set that defines a system or sets the conditions of its operation.

ABSTRACT

Parameterization is the task of quantifying the static values needed by the algorithms in the selected model. Parameters may be equation coefficients, threshold values, and vegetation categories. Parameterization is a preliminary modeling phase that is absolutely required in all modeling projects because it will not only quantify the information needed to run the model, but it will also help the user understand the model and the algorithms that comprise the model. First, this chapter covers the five main methods of quantifying parameters—field measurements, legacy data, literature, model approximations, and expert opinion. Then, the chapter covers how to design a comprehensive parameterization strategy to ensure the most appropriate values are assigned to all parameters. Then, the seven steps in a parameterization are presented: (1) develop strategy, (2) identify parameters, (3) collect data, (4) quantify parameters, (5) run model, (6) evaluate results, and (7) revise parameters. And last, various aspects of performing a successful parameterization are covered.

Introduction

Parameterization is the task of quantifying the static values (i.e., parameters) needed by the algorithms in the model that mathematically represent simulated ecological processes. Some parameters are coefficients to equations (e.g., light compensation point), while others are threshold values that are used to decide a specialized procedure or algorithm in the model (e.g., cover of woody species). Some parameters are solely at the discretion of the user, such as the values that set simulation design (e.g., number of years to simulate, types of outputs, and size of pixels), and others are specific to the simulated module or ecosystem process (e.g., maximum leaf area, soil water holding capacity). Parameters are the heart of the model and represent the best way to tailor a model for a specific local application.

Parameterization is a preliminary modeling phase that is absolutely required in all modeling projects (Grimm and Railsback 2013). This phase will not only quantify the information needed to run the model, but it will also help the user understand the model and the algorithms that comprise the model. Poor parameterizations are major problems in many modeling projects (i.e., garbage in, garbage out), so sufficient time should be dedicated to this effort (Jakeman et al. 2006). Unfortunately, it is also the phase when most users feel that they are "out of their element" —many of the parameters and their meanings might be unknown to users unless they are familiar with the literature. Hopefully this chapter will help novice users through this often demanding phase.

There is never a perfect parameterization, except for the simplest of ecological models, which are rarely used in land management activities.

It is difficult, if not impossible, to flawlessly parameterize mechanistic models because of their inherent complexity; compromises are always made in quantifying model parameters because of measurement limitations, scale, paucity of data, and context incompatibilities and inconsistencies (McCallum 2008). Because of this, users often become discouraged and disappointed with their parameterizations. But a disappointing parameterization is never a reason to abandon all confidence in model output; other phases of the modeling project, specifically calibration and validation, often help to address limitations of the parameterization and provide ways to improve the parameterization. And, the uncertainty can always be incorporated in the analysis and interpretation of model findings.

Parameterization Methods

There are five major methods that are often used to quantify model parameters and they involve estimating parameters from the following sources (in order of preference): field measurements, existing data syntheses, literature searches, iterative model approximations, and expert opinions. Each of these methods has advantages and disadvantages that must be addressed under any project design (Table 5.1). If time is short, for example, it may be impossible to collect field measurements or conduct model approximations. Most model parametrizations employ a blending of all methods to best estimate the diverse parameters that can be part of applying complex models, such as LESMs.

Field measurements

Field measurements within the geographic area and environmental context of the modeling project can provide more accurate parameter values than nearly all other sources. Site, water, plant, and animal characteristics, for example, can be measured in the field to quantify parameters in the model. However, some parameters in the model may be difficult to measure because they require specialized equipment and/or demand inordinate time and resources. As examples, many forest gap models require the maximum height and diameter for each tree species to parameterize the growth algorithm, and these characteristics may be difficult to measure over a

Table 5.1. A summary of the various methods that can be used to quantify a model's parameter. The methods are listed in order of preference.

Method	Advantages	Disadvantages
Field Sampling	Best method for estimating parameters; can target specific parameters; can be also be used to sample for initialization, calibration, validation.	Impractical in most projects because of lack of time, funds, or experience.
Existing Data Compilations	Popular; simple; can be localized; often best source for most projects.	Local data sources difficult to find; often need to reformat data; requires knowledge of data systems; may require specialized knowledge and high computing power to process data; limited for most areas; may have scale and organization inconsistencies.
Literature Searches	Most common; easy; abundant information.	Time-consuming; Never really finished; Often confusing and contradictory; Local studies difficult to find; Study measurements inappropriate to project's location or conditions; units are often wrong and difficult to convert.
Model Approximations	Provides a safety net in case no other method possible; can be simple for some parameters.	Difficult to perform for some parameters; Requires knowledge of model or computers; Only possible for a limited number of parameters.
Expert Evaluations	Quick; easy; review other parameters at same time.	High degree of uncertainty; often disagreement among experts; High subjectivity because only one or two experts contribute.

limited seasonal field campaign (Botkin 1993). Many evapotranspiration and photosynthesis algorithms need highly specialized parameters, such as the light compensation point, maximum stomatal conductance, and temperature thresholds, that must be measured by expensive equipment such as gas-exchange systems (Hari et al. 1990). And many fire models need an estimate of fire return intervals (FRIs) for a given landscape and estimating historic FRIs involves intensive sampling of fire-scars taken from live and dead trees and bringing the fire-scar samples into the lab for analyses (Maruoka and Agee 1994). Since field measurements are costly and time-consuming, field sampling is only practical for model parameterization when there are sufficient resources. However, if field sampling is possible, users should be sure to read the important suggestions from Chapter 3 that might make that effort more effective (summarize here):

1. Merge data collection with the three previously mentioned phases (initialization, calibration, and validation).
2. Ensure field data are useful to your organization after modeling.
3. Collect data that may be used to understand the parameter.
4. Use methods and protocols that mesh with organizational requirements.
5. Confirm that the sampled data can be used for parameterization.

As mentioned in the previous chapter, there are a number of standardized methodologies that could be used as templates for field sampling in parameterization field campaigns, such as FIREMON (Lutes et al. 2006) and FFI (Lutes et al. 2009). However, these packages will have limited use in measuring the diverse and specialized parameters found in complex models. Perhaps the best way to obtain sampling procedures for the specialized and exacting parameters is to find those published papers where the parameters were quantified by others.

Existing data compilation

There should be a comprehensive search for existing data to use for parametrization, and this search should be both extensive and flexible. First start within the home organization or agency to find unpublished data from reports and local databases that might be useful for parameterization. Then, identify any national efforts that collected appropriate field data that may be useful for parameterization. National programs such as the US Forest Service's FIA (Bechtold and Patterson 2005), USGS grassland inventories (Council 1994), and the NSF NEON project (Senkowsky 2003) might have suitable data with easy access. Next, universities, government agencies, NGO's and other potential partners that might have relevant data should be contacted. Global databases may also be a source of parameter data such, as the TRY Plant Trait Database, but may require requests for access (Kattge et al. 2011).

Remotely sensed data (e.g., satellite imagery, Lidar, radar) might also provide a way of estimating some model parameters. Lidar products, for example, can be used to estimate canopy fuel characteristics (Andersen et al. 2005, Erdody and Moskal 2010), canopy cover and leaf area (Korhonen et al. 2011), and forest and tree structure and height (Lim et al. 2003). And ground-based Lidar can be used to estimate various characteristics of the forest floor (Loudermilk et al. 2009). Passive sensors, such as Landsat (e.g., TM, ETM, OLI), MODIS, Sentinel, and AVHRR, can be used to estimate phenological parameters (Ehrlich et al. 1994, Zhang et al. 2003, Delbart et al. 2006). Land cover and land use are often mapped and monitored using a wide variety of remote sensing platforms (Rogan and Chen 2004). However, the effective use of these spatial data resources may demand a high level of expertise from the user or project team.

Literature searches

The next alternative for quantifying parameters is to peruse the literature to glean values for parameters from published studies (White et al. 2000). This is easily the most popular approach among users because of the paucity of existing data to quantify important biophysical parameters. Especially precious are published synthesis papers that have compiled extensive summaries of some important modeling parameters; Hessl et al. (2004), as an example, published a set of ecophysiological parameters synthesized from the literature for the tree species of the Pacific Northwest USA. If these compilations are unavailable, then each parameter must be researched independently. Users should begin their search in the modeling literature starting with papers written for their selected model, as most modeling papers publish the list of parameters used in the modeling study. Then users should expand the search to similar models that may have used the same parameter. If fruitless, then users can move to the empirical literature that concerns field studies of the desired parameter and its estimation. Respiration coefficients, for example, could be published in the ecophysiological literature. And last, users should ask local experts at universities and research stations if they know of any literature or unpublished work that would help your parameterization.

Users should also employ creative methods to find parameters in the literature. Some papers provide correlations of more easily measured variables to the difficult model parameters, and these regression equations can be used to estimate values for their project. Often, there will be another variable that is measured in the literature that can be easily converted to approximate a value for the desired parameter; canopy cover (%), for example, can be used to coarsely estimate leaf area (Carlson and Ripley 1997). Users can also query the author(s) of related papers asking them if they have any information on the desired parameter, such as from unpublished studies, and they may also contact author(s) of similar models to ask them to send input parameter files. Users should know that the literature will contain many contradictory and confusing values for some parameters. Sometimes, the range of values for a single parameter will be double its mean because it was measured for a different place, for a different species, or for a different time, or even with different equipment.

Abundant notes should be taken concerning details of literature searches. Values from the literature can be recorded in spreadsheets and the sources can be meticulously described along with any possible notes considering measurement of the parameter. The electronic copy of the paper is usually stored in a well-designed directory structure and the reference is imported into citation software or an electronic document. A well-documented parameterization will be invaluable to others who plan to use the model, especially those within the home organization of the user, and it will also help in the analysis and report-writing. It is important to record whether the discovered parameter values are appropriate for your specific ecosystem or location. Often only single values may be reported for a parameter, but if it was measured in a different place, under different conditions, or following a different protocol you may not be able to use the value for your model with confidence.

Model approximations

Another less used, but often invaluable approach, is the use of the model itself to estimate the parameter's value (Jakeman et al. 2006). This approach is only appropriate when there are ample field measurements on the algorithm in which the parameter is used. If a parameter's value is completely unknown, and there are few data or limited literature studies to help in its quantification, the user can make an educated guess as to the initial value of the parameter. That initial value is then entered into the model and the model is executed and results from the algorithm in which the parameter occurs are compared against real observations. This process is quite similar to that used in the calibration procedure (Chapter 6). If the resultant values disagree with observations, the parameter's value is then altered and the model is run again. This procedure is repeated until the user is satisfied that the revised value of the parameter is providing realistic results. An alternative to this method is to use another, more focused, model or computer program to estimate the parameter. For example, the algorithm can be separately programmed for ease of use for model parameterization. Or, sometimes others have built computer programs specifically for estimating a particular parameter; Hill (1992), for example, created a model to estimate parameters in a transient ground water flow model and Kruuk (2004) estimated genetic parameters using maximum likelihood statistical models.

This iterative method assumes several things. First, all the other parameters in the algorithm are reasonably estimated. Second, the processes in the algorithm are accurately represented and programmed correctly. Next, the measurement of the reference observations that the user is using for comparison are consistent with modeled output (e.g., correct units). And last, the user has the experience and knowledge to iteratively run the model. Another variation to this method is to use experts instead of measured observations as reference in the evaluation. This model approximation method is time-consuming and quite demanding, especially if there a many parameters that need quantification, therefore many modelers use this as a last resort.

Expert evaluation

The last alternative is using expert opinion for model parameterization. In this approach, a panel of experts is employed to estimate parameter values. Invariability, experts often need some knowledge of the model and its behavior before they provide estimates, so it is important that the user provide a comprehensive package for the experts to review including

101

results of various model runs showing the influence of a gradient of parameter values (see Chapter 6). The same expert team can also be used to review the parameters after this parameterization phase is complete so the user can gain additional confidence in model results. Obviously, using experts to estimate parameters is less desirable than the other methods because of its inherent subjectivity and human biases, but many users will reluctantly find that this is an indispensable tool in parameterization.

Parameterization strategy

Because users often find that some model parameters are difficult to quantify, it is strongly suggested that the user develop a plan for dealing with inadequate and inconsistent information to quantify parameters. A hierarchical decision-making process can be developed to quantify those parameters whose values are unknown or suspect so that bias is kept to a minimum. During a parameterization, for example, there are many parameters that cannot be quantified because their values were never measured or there was little time to measure them for the project, especially for complex models. The question then is what to do about the missing parameter values? Do you select another model or try to find an alternative way to quantify missing parameters? The user should design a comprehensive plan that is employed every time an un-measured parameter is encountered. This will mitigate some of the subjectivity that can creep into a parameterization and it also will make the process more efficient. The hierarchy of selecting the method used to estimate a parameter's value is the first thing to explicitly document in a parameterization effort. As an example, to quantify a parameter, values from field measurements might be selected first, then values from the literature are used if no field values are available, then values from existing data if no literature values, then experts, then best "guess", and last model approximation. Or there may be some overriding criteria that need to be followed, such as proximity to simulation area, use in similar models, and rigor in measurement or estimation. At any rate, the melding of all approaches in a comprehensive plan is the best way to estimate parameter values.

If insufficient data exists for a parameter of a given species, one parameterization strategy is for the user to assign the same value as that of another species of the same genus (e.g., the same parameter can be used for all *Pinus* genera). If the entire genera are missing parameters, then a parameter from species of the same plant functional type (PFT) (Smith et al. 1993) can be used. PFTs have been used by many modelers and ecologists as a way to synthesize species characteristics (Diaz and Cabido 1997, Enright et al. 2014, Stahl et al. 2014). PFTs are also called plant guilds (Wilson 1999), functional groups, and functional classifications (Gitay and Noble 1997). Species are often grouped together based on their representation in the model, such as by ecophysiological characteristics (Thonicke et al. 2001), response to disturbance (Lavorel et al. 1997), life histories (Adler et al. 2014), successional dynamics (Noble and Gitay 1996), and land use (Gondard et al. 2003). PFTs can be a powerful tool in other phases of the modeling project, such as validation (e.g., summarizing model results across PFTs), calibration (e.g., using PFTs as a way to stratify diagnostic variables), and initialization (e.g., assigning missing age data based on age distributions within a PFT).

Perhaps the best way to illustrate a parameterization strategy is to provide a simple example. In a parameterization of a complex model, such as an LESM, a user might find that direct measurement of parameters in the growth algorithms for some tree species is impractical because of time and cost limitations; only existing unpublished data and data from the literature, in that order, are possible within the project's timeline and budget (e.g., no time or funds to perform a field campaign, or model approximation effort). Therefore the user might decide that any species that are missing parameter values will receive the same value as that of other species of the same genus that have a value (e.g., the same parameter can be used for all *Pinus* genera). If the parameter value is missing for the entire genera, then a parameter from species of the same plant functional type (PFT) (Smith et al. 1993) might be used. The PFTs could be based on lifeform (conifer, broadleaf, shrub, and herb). If there are no genera or PFT parameter values for a certain species, then the parameter could be estimated as an average across all species. And as a last resort, the parameter is estimated from expert opinion. If multiple values of the same parameter are found in the literature for the same species, then the one that was measured closest to the simulation landscape could be the best option; the one that was measured directly or measured with the best procedures might be taken next; and last, an average across all parameter values could be used. Keane et al. (1996) used a similar form of this approach in the first parameterization of a complex LESM. This plan ensured consistency in parameter estimation and reduced possible subjectivity that may be encountered in the parameterization process. But most importantly, this plan provided a means of communication across the modeling team. A more complex strategy is presented as an example in Table 5.2.

A critical part of developing a parameterization plan is to document everything and provide ample detail describing each step. The value(s) of each parameter must be documented in the context of the simulation plan. Parameter estimations that use multiple methods will also need to be detailed. This information will be essential in several other project phases. Here's a hypothetical example to illustrate the process of documentation. A simple model required a set of ecophysiological parameters for simulating photosynthesis (light compensation point), respiration (Q10 coefficient), and evapotranspiration (stomatal conductance) for multiple species. The user created a spreadsheet of the parameters (columns) by species (rows) and populated the spreadsheet with values measured in the field, found in the literature, and obtained using model iteration, in that order. The sources used for parameter estimation were documented on another spreadsheet as well as any other interpretive notes. However, about 10% of the parameters had missing values, so the user decided to employ PFTs used in the modeling literature (Running et al. 1994, Adler et al. 2014) as a way to assign missing values. The PFTs were designed to distinguish ecophysiological characteristics using criteria such as shade tolerance, leaf morphology, drought tolerance, and disturbance adaptations. The user then grouped species implemented in the model into PFTs and assigned missing parameter values across the PFT classification, documenting every step in the spreadsheet.

Table 5.2. An example 10-level parameterization strategy for estimating the values of parameters for a plant species in a simple ecological model. Pretend the model is a spatial gap model that has 20 tree species included in its architecture and there are four parameters being estimated for each species: maximum DBH, maximum height, maximum age, and shade tolerance level (1-low, 5-high). The user would use the following hierarchy to quantify each of the four parameters. The level would then be recorded in the spreadsheet used to store the parameter values. This spreadsheet may have more than one column for each value in case multiple values are available in existing databases, in the literature, and from experts.

Levels	Procedure	Notes
I	Estimate from field data collected specifically for the project at or near the area to be simulated	Field campaigns can be used to sample one or more of the four parameters
II	Estimate from field data collected by someone else and published in the peer-reviewed literature that is NEAR the landscape	Peruse the literature and determine if there were studies near the simulation landscape that measured the parameters
III	Estimate from field data collected by someone else and published in the peer-reviewed literature that is DISTANT from the landscape	Be sure to set guidelines on how to determine near and far
IV	Estimate from field data collected by someone else NEAR the landscape but remains unpublished	Inventory all existing databases to see if anyone sampled the parameter(s)
V	Estimate from field data collected by someone else DISTANT from the landscape but remains unpublished	Be sure to use the same guidelines for near and far as above
VI	Approximate using model iterations	Use the model to iteratively estimate the parameter values
VII	Obtain a value from experts	Ask local experts (e.g., modelers, foresters, biologists) for values
VIII	Use a value from a species of the same genus	Be sure to record at what level the original value was approximated
IX	Use a value from a species of the same plant functional type	Must develop a PFT classification; could use one from the literature; be sure to record original level
X	Estimate the value	Use your own expertise in the model to approximate a value for a species

Parameterization Steps

The steps involved in model parameterization can be synthesized into six primary actions (Figure 5.1). In this section, these six steps are outlined to illustrate an effective parameterization process for any modeling project.

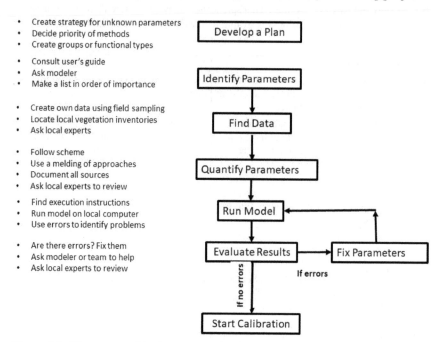

- Create strategy for unknown parameters
- Decide priority of methods
- Create groups or functional types

- Consult user's guide
- Ask modeler
- Make a list in order of importance

- Create own data using field sampling
- Locate local vegetation inventories
- Ask local experts

- Follow scheme
- Use a melding of approaches
- Document all sources
- Ask local experts to review

- Find execution instructions
- Run model on local computer
- Use errors to identify problems

- Are there errors? Fix them
- Ask modeler or team to help
- Ask local experts to review

Develop a Plan → Identify Parameters → Find Data → Quantify Parameters → Run Model → Evaluate Results → Fix Parameters

If errors / If no errors → Start Calibration

Figure 5.1. The steps used in the parameterization phase of a modeling project. Steps in italics indicate that the step is optional. Various tasks are summarized to the left of each step.

Develop a parameterization strategy

This initial step in any parameterization, which is the creation of a parameterization plan as discussed above, will form the foundation on which to build a parametrization and provide a structure to make parameterization decisions. It is easily the most critical step because all decisions can be made in this context. Creation of the plan should use material presented in the previous section, along with any local knowledge of the modeling project that may be important in parameterization. The plan should be thoroughly documented and revised as new situations are encountered during parameterization.

Identify parameters

In this step, the parameters that need quantification are identified using all reference material for the selected model. Some model parameters are relatively simple and their quantification can be done easily. For example, most simulation parameters (output variables, simulation time span, years in the weather record) are estimated in the context of the simulation design. However, parameters in key algorithms of the model may require more effort because they must be carefully estimated. When there are abundant parameters that need estimation, it is often beneficial to rate the parameters as to their importance in the simulation architecture or their influence on model output. Past sensitivity analysis results (Chapter 8), previous modeling projects, information from the literature or model manuals, objectives of the modeling project, and basic ecological knowledge can be used to rate model parameters. The most important parameters should receive the greatest time and effort in their estimation.

Find data

All available data should be identified for quantifying model parameters. As mentioned in the previous section, data sources can be from field sampling efforts, existing data, the literature, model approximations and experts. Users may find that they use all these methods to estimate values for the diversity of model parameters.

Quantify parameters

In this step, all available data and knowledge are used to quantify values for all parameters identified by the user. One or more of the above methods are used to estimate the parameter values, and often, a melding of methods is the most effective approach in the estimation of all parameters. It is critically important that the sources used for parameter estimation are thoroughly documented because this will help in the calibration, validation, and analysis phases. Here is where the user implements the parameterization strategy to quantify parameters, which may involve prioritizing the methods used to assign a value; protocols to use when information on a parameter is unknown, the hierarchical ordering of procedures to use when more than one value exists for a parameter; and venues of documentation (e.g., spreadsheets, illustrations, modeling journal).

Run model

Once all parameters have been estimated, they are entered into the input files of the model or directly into the model, and the model is run to ensure that each parameter has been properly entered. Many modeling projects were stalled because the model improperly read the parameter value because of various circumstances, such as the wrong format, units, or decimal point.

Evaluate results

Model output should be closely examined to determine if the parameters were input correctly and were appropriately quantified. The first evaluation is to detect any model error or warning messages that occurred because of the improper parameterization. Next, model output should be examined to ensure that the parameter is being properly accepted by the model. Many models produce a file documenting all the parameters as they were read by the computer program (e.g., "echo" files) and these output files are invaluable for debugging input files. Otherwise, the user is left with checking simulated results from a particular module to determine if they are believable. If this is the only alternative, then the user should evaluate the entire parameterization rather than evaluate one parameter at a time. If results appear to be in error, then the parameters need to be revised as they were probably improperly entered.

Fix parameters

Fix all erroneous parameters and rerun the model until model output seems to reasonably represent ecological processes of interest in a landscape. This might be a good time to ensure that the units on all the parameter values are correct; many a modeling project was waylaid for weeks because of a simple error in a major parameter value's units. Also, another useful error check is to examine the magnitude of model outputs. For example, some variables output by the model may be orders of magnitude higher or lower than from expected values due to incorrect parameterization.

Parameterization Concerns

Although model parameters are the core of the simulation effort and they provide the only way a model can be adapted to local situations, values for these important model inputs are sometimes extremely elusive. As mentioned, the lack of comprehensive assessments of parameters often leads to dissatisfaction and disappointment in model users. But the user must remember that there is a great deal of uncertainty in the entire modeling process and a "perfect" parameterization may never be possible. Moreover, spending time to obtain a parameterization that the user feels is sufficient may have not appreciably improve model results. A photosynthesis algorithm, for example, may have so many internal sources of variation that a perfect estimate of one parameter may do little to mitigate the underlying uncertainty in algorithm development. Or, an algorithm may be driven by one key parameter, while others in the equation have a relatively minor influence on model estimates. However, a thoughtful parameterization can provide confidence in model predictions, and it may also allow the user to more thoroughly understand the model. In the end, the user should take comfort in knowing that parameterization was done using the best available scientific information in a timeline appropriate for the project. If there are serious doubts about the parameterization, they can be addressed in the interpretation of the model's results.

In my experience, a good parameterization is often limited by the timelines of the project rather than the availability of the possible data sources to quantify parameters. Literature searches for specialized ecophysiological parameters, for example, may take months to complete and, in fact, are rarely ever finished because new research is always being published. In addition, the compilation and analysis of newly published data to quantify parameters may take inordinate amounts of time. It is often best to select a "sunset" date for the parameterization phase and decide, on that date, if the parameterization is acceptable. If not, project timelines will need to be redone based on educated guesses on how much longer it will take to parameterize the model to an acceptable level. Obviously, it is best to involve the modeler in the parameterization to ensure the final product is acceptable. It's also beneficial to have a group of experts review the parameter set after its initial quantification. The scientific and modeling community may have some ideas of how best to parameterize a model and if the parameters estimates are acceptable. This will provide additional confidence in the parameterization.

And last, it is important that the user never perform the evaluation of the initial conditions (Chapter 4) and parameterization at the same time because it may be difficult to determine which values are in error if the model performs poorly. Conduct each task separately, and perform the initialization evaluation first.

References

Adler, P. B., R. Salguero-Gómez, A. Compagnoni, J. S. Hsu, J. Ray-Mukherjee, C. Mbeau-Ache and M. Franco. 2014. Functional traits explain variation in plant life history strategies. Proceedings of the National Academy of Sciences 111: 740–745.

Andersen, H.-E., R. J. McGaughey and S. E. Reutebuch. 2005. Estimating forest canopy fuel parameters using LIDAR data. Remote Sensing of Environment 94: 441–449.

Bechtold, W. A. and P. L. Patterson. 2005. The enhanced forest inventory and analysis program-national sampling design and estimation procedures. Gen. Tech. Rep. SRS-80. Asheville, NC: US Department of Agriculture, Forest Service, Southern Research Station. 85 p. 80.

Botkin, D. B. 1993. Forest Dynamics: An Ecological Model. Oxford University Press., New York, NY., USA.

Carlson, T. N. and D. A. Ripley. 1997. On the relation between NDVI, Fractional vegetation cover, and leaf area index. Remote Sensing of the Environment 62: 241–252.

Council, N. R. 1994. Rangeland health: New methods to classify, inventory, and monitor rangelands. National Academies Press.

Delbart, N., T. Le Toan, L. Kergoat, and V. Fedotova. 2006. Remote sensing of spring phenology in boreal regions: A free of snow-effect method using NOAA-AVHRR and SPOT-VGT data (1982–2004). Remote Sensing of Environment 101: 52–62.

Diaz, S. and M. Cabido. 1997. Plant functional types and ecosystem function in relation to global change. Journal of Vegetation Science 8: 121–133.

Ehrlich, D., J. E. Estes and A. Singh. 1994. Applications of NOAA-AVHRR 1 km data for environmental monitoring. International Journal of Remote Sensing 15: 145–161.

Enright, N. J., J. B. Fontaine, B. B. Lamont, B. P. Miller and V. C. Westcott. 2014. Resistance and resilience to changing climate and fire regime depend on plant functional traits. Journal of Ecology 102: 1572–1581.

Erdody, T. L. and L. M. Moskal. 2010. Fusion of LiDAR and imagery for estimating forest canopy fuels. Remote Sensing of Environment 114: 725–737.

Gitay, H. and I. Noble. 1997. What are functional types and how should we seek them. Plant functional types: Their relevance to ecosystem properties and global change 1: 3–19.

Gondard, H., S. Jauffret, J. Aronson and S. Lavorel. 2003. Plant functional types: A promising tool for management and restoration of degraded lands. Applied Vegetation Science 6: 223–234.

Grimm, V. and S. F. Railsback. 2013. Individual-Based Modeling and Ecology. Princeton University Press.

Hari, P., E. Korpilahti, T. Pohja and P. K. Räsänen. 1990. A field system for measuring the gas exchange of forest trees. Silva Fennica 24: 21–27.

Hessl, A. E., C. Milesi, M. A. White, D. L. Peterson and R. E. Keane. 2004. Ecophysiological parameters for Pacific Northwest trees. General Technical Report PNW-GTR-618, USDA Forest Service Pacific Northwest Research Station, Portland, OR, USA.

Hill, M. C. 1992. A computer program (MODFLOWP) for estimating parameters of a transient, three-dimensional ground-water flow model using nonlinear regression. 2331–1258, US Geological Survey.

Jakeman, A. J., R. A. Letcher and J. P. Norton. 2006. Ten iterative steps in development and evaluation of environmental models. Environmental Modelling & Software 21: 602–614.

Kattge, J., S. Diaz, S. Lavorel, I. C. Prentice, P. Leadley, G. Bönisch, E. Garnier, M. Westoby, P. B. Reich and I. J. Wright. 2011. TRY—a global database of plant traits. Global Change Biology 17: 2905–2935.

Keane, R. E., P. Morgan and S. W. Running. 1996. FIRE-BGC—a mechanistic ecological process model for simulating fire succession on coniferous forest landscapes of the northern Rocky Mountains. Research Paper INT-RP-484, United States Department of Agriculture, Forest Service Intermountain Forest and Range Experiment Station, Ogden, UT USA.

Korhonen, L., I. Korpela, J. Heiskanen and M. Maltamo. 2011. Airborne discrete-return LIDAR data in the estimation of vertical canopy cover, angular canopy closure and leaf area index. Remote Sensing of Environment 115: 1065–1080.

Kruuk, L. E. 2004. Estimating genetic parameters in natural populations using the 'animal model'. Philosophical Transactions of the Royal Society of London B: Biological Sciences 359: 873–890.

Lavorel, S., S. McIntyre, J. Landsberg and T. D. A. Forbes. 1997. Plant functional classifications: From general groups to specific groups based on response to disturbance. Trends in Ecology & Evolution 12: 474–478.

Lim, K., P. Treitz, M. Wulder, B. St-Onge and M. Flood. 2003. LiDAR remote sensing of forest structure. Progress in Physical Geography 27: 88–106.

Loudermilk, E. L., J. K. Hiers, J. J. O'Brien, R. J. Mitchell, A. Singhania, J. C. Fernandez, W. P. Cropper and K. C. Slatton. 2009. Ground-based LIDAR: a novel approach to quantify fine-scale fuelbed characteristics. International Journal of Wildland Fire 18: 676–685.

Lutes, D. C., R. E. Keane, J. F. Caratti, C. H. Key, N. C. Benson, S. Sutherland and L. J. Gangi. 2006. FIREMON: Fire effects monitoring and inventory system. General Technical Report RMRS-GTR-164-CD, USDA Forest Service Rocky Mountain Research Station, Fort Collins, CO USA.

Lutes, D. C., N. C. Benson, M. Keifer, J. F. Caratti and S. A. Streetman. 2009. FFI: a software tool for ecological monitoring*. International Journal of Wildland Fire 18: 310–314.

Maruoka, K. R. and J. K. Agee. 1994. Fire Histories: Overview of Methods and Applications. Technical Notes BMNRI-TN-2, Blue Mountains Natural Resources Institute.

McCallum, H. 2008. Population Parameters: Estimation for Ecological Models. John Wiley & Sons.

Noble, I. R. and H. Gitay. 1996. A functional classification for predicting the dynamics of landscapes. Journal of Vegetation Science 7: 329–336.

Rogan, J., and D. Chen. 2004. Remote sensing technology for mapping and monitoring land-cover and land-use change. Progress in planning 61: 301–325.

Running, S. W., T. R. Loveland and L. L. Pierce. 1994. A vegetation classification logic based on remote sensing for use in global biogeochemical models. Ambio 23: 77–81.

Senkowsky, S. 2003. NEON: Planning for a New Frontier in Biology. BioScience 53: 456–461.

Smith, T., H. Shugart, F. Woodward and P. Burton. 1993. Plant functional types. pp. 272–292. Vegetation Dynamics & Global Change. Springer.

Stahl, U., B. Reu and C. Wirth. 2014. Predicting species' range limits from functional traits for the tree flora of North America. Proceedings of the National Academy of Sciences 111: 13739–13744.

Thonicke, K., S. Venevsky, S. Sitch and W. Cramer. 2001. The role of fire disturbance for global vegetation dynamics: Coupling fire into a Dynamic Global Vegetation Model. Global Ecology and Biogeography 10: 661–677.

White, M., A., P. Thornton, E., S. W. Running and R. Nemani. 2000. Parameterization and sensitivity analysis of the Biome-BGC terrestrial ecosystem model: Net primary production controls. Earth Interactions 4: 1–85.

Wilson, J. B. 1999. Guilds, functional types and ecological groups. Oikos: 507–522.

Zhang, X., M. A. Friedl, C. B. Schaaf, A. H. Strahler, J. C. Hodges, F. Gao, B. C. Reed and A. Huete. 2003. Monitoring vegetation phenology using MODIS. Remote Sensing of Environment 84: 471–475.

6

Calibration

How to Tune the Model for Realism

"Our abilities to understand, adjust, and change make us wise."

Debasish Mridha M.D.

Calibration—to plan, adjust, or devise carefully to have a precise use or application.

ABSTRACT

Calibration is the process of evaluating preliminary model output and adjusting parameters to get a more realistic result. Calibration performs many important tasks—the user (1) develops a solid feel for model behavior from the iterative runs, (2) becomes more confident in interpreting results, (3) detects obvious problems with initialization and parameterization, (4) learns the sensitivities of model parameters, and (5) prepares the model for the execution of the project. This chapter presents the eight steps of calibration: (1) identify important parameters to use as knobs for calibration, (2) develop a calibration strategy to ensure objectivity, (3) collect any data or information useful in the calibration, (4) perform preliminary calibration simulations, (5) compile a final list of parameters, (6) adjust parameters, (7) run model, and (8) evaluate results and repeat. Last, a set of issues that emerge during calibration are addressed and solutions are presented.

Introduction

Calibration, also known as 'conditional verification' (Jakeman et al. 2006), is the process of evaluating preliminary model output for realism and believability, and if results seem unrealistic, then adjusting parameters to achieve a higher level of realism (Rykiel 1996). This is often an exhausting and perplexing task, especially for novice users of complex models. It is exhausting because it employs a classic trial-and-error or iterative method of improving believability of model results by comparing them with existing information and changing the appropriate parameters to get better comparisons. It is perplexing because often the user has limited *a priori* knowledge of the model, influences of parameters, and model behavior, making the entire task somewhat of an paradox. Calibration can take substantial time, especially for large, complex models, such as LESMs, because complex models generate profuse results for many diverse variables, each of which may need to be evaluated for realism. Moreover, changing one parameter may cause output for one variable to become more realistic or acceptable, but it may also may result in another variable's values being less believable. The model must be rerun many times to get to a point where the user feels that the model has sufficient realism and is ready for validation and execution phases in their specific project. Through this process, however, the user of the model gains a better understanding of the model and why it behaves the way it does, and it is the calibration process where the user develops the deep understanding needed to interpret the final results of the modeling effort.

Calibration is different from validation (Chapter 7) in that, in calibration, the initial conditions and parameters of the model are changed to get better, more believable model results, whereas in validation, the model results are evaluated against actual data to determine accuracy and precision and the model is NOT modified based on the validation. Moreover, calibration should be done before the model is validated.

Calibration is needed any time a model is applied to a new area or used in a new way because input parameters are often replaced or adjusted for the new situation (Janssen and Heuberger 1995). Parameters value are rarely perfect fits in the selected model for a variety of reasons. Their values may have been estimated rather than measured, or their values may have been extrapolated from a different situation and therefore may not be compatible with the current use. For example, parameters that drive a plant respiration algorithm may have been measured for a plant species, size, or condition that are different from what is simulated in the selected model. Moreover, these parameters may have been measured for a single plant under greenhouse conditions whereas the model may need to simulate an entire array of interacting plants in the natural environment. As a result, measured parameter values may not always be exactly what is needed by the model. Adjustments must be made, using the model as a sounding board, to ensure parameters are realistic and acceptable.

It may seem to many that calibration is "fudging" model results to get what the user wants for an answer or what the user believes to be the desired result. However, calibration is actually a little more involved than this simplistic perception. Calibration performs many important tasks. First, it allows the user to get a solid feel for model behavior from the iterative runs, and as a result, the user becomes more confident in interpreting results during the analysis phase of the modeling project. Second, calibration often detects obvious problems with initialization and parameterization, and those inconsistencies can be dealt with prior to model execution. A common problem encountered by many landscape modelers is the model is run for a century or two without disturbance and the shade tolerant species fail to outcompete the shade intolerant species because of inappropriate ecophysiological parameter values for some species. The sensitivities of some model parameters also become evident during the calibration stage; modification of one parameter may result in small changes in model results, and as such, may be a minor source of error in simulations.

It should be obvious that the process of calibration has a major challenge —how is the user able to determine realism and believability if they acknowledged that they needed the model to understand what will happen in the future. However, calibration is more than addressing response variables only; it involves evaluating other diagnostic variables that are important influences on response variables. For example, say the modeling objective is to determine changes in tree species composition across two scenarios. Users may evaluate the response variable of tree density by species, but they should also evaluate disturbance, climate, and plant size diagnostic variables to ensure the model is acting properly and generating acceptable results. Output variables such as annual area burned by wildfire, average fire return interval, number trees killed by insects, potential evapotranspiration, and diameter of living trees might be included in the inspection for realism. These diagnostic variables are like the gauges on a car's dashboard and calibration is the process of adjusting engine settings to ensure the needles on the gauges are at the right place. The engine settings are the parameters that were quantified for the modeling project (Chapter 5). Modelers often refer to these important parameters as "knobs" that can be adjusted to modify model behavior (i.e., the needle). Indeed, some parameters are so important to the simulation that small adjustments can lead to dramatic changes in simulation results (i.e., sensitive—see Chapter 7). So, even though it may appear that users are adjusting parameters to get desired results, calibration is a necessary task, and if done systematically with a comprehensive and objective strategy and suitable resource materials, it can be a rewarding and critical step in the modeling project.

There are seven steps involved in the calibration phase (Figure 6.1). The first four build on the last three, which are iterative. Users can calibrate all or parts of the parameterization input data set depending on time and resources. The first two of these steps can be done simultaneously.

An example is used throughout this chapter that might help illustrate the power of a well-designed calibration strategy. As background for this example, we propose that initialization and parameterization were completed for a landscape modeling project whose objective was to evaluate changes in basal area over 50 years under four management scenarios. The calibration phase is ready to begin. The selected model was a mechanistic gap model implemented in a landscape context. The modeling team was composed

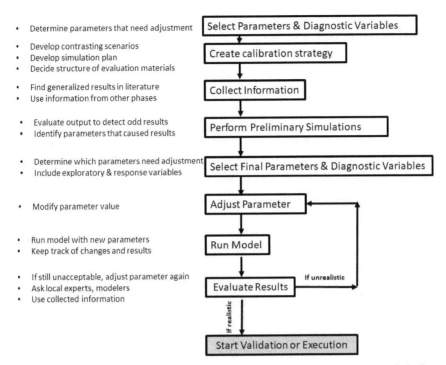

Figure 6.1. The steps used in the calibration phase of a modeling project. Steps in italics indicate that the step is optional. Various tasks are summarized to the left of each step.

of a wide variety of resource professionals, including a hydrologist, fire management specialist, and two ecologists, but missing from the team were foresters who would actually understand basal area dynamics. The primary user, one of the ecologists, spent many years evaluating species interactions, especially involving trees, climate, and wildlife. There was ample information in the literature and from local studies on the historical vegetation and disturbance dynamics of the landscape. The modeling team designed four contrasting scenarios to comprehensively evaluate model behavior and calibrate accordingly. The first scenario was designed to simulate historical landscape dynamics using historical weather and wildland fire parameters because there were ample information in the literature to compare with the output. The second scenario was designed to simulate landscape dynamics in the absence of disturbance, specifically

117

fire, because this would therefore highlight successional dynamics, an topic area well known by the primary user, an ecologist. A third scenario was designed to simulate historical landscape dynamics to replicate some semblance of a fire record that could then be assessed by the fire specialist. This scenario could also be evaluated by others on the modeling team under the umbrella of their own field of expertise. And last, the model would be run with a set of thinning management actions implemented on special areas and at special times within the landscape in the absence of any disturbance. This last scenario would allow for the evaluation of management-oriented parameters. The modeling team determined that to adequately capture landscape dynamics, the model would need to be run for at least 50 years and that results would need to be printed every 5 years across each scenario. For diagnostic variables, the basal area response variable was selected along with the explanatory variables of tree density and average DBH across all species, the area burned by fire severity level, and the timber volume harvested. The team also decided to add potential evapotranspiration, density of trees killed by insects and disease, and average soil water potential as other key diagnostic variables because of existing expertise on the team. The team also decided that the best way to evaluate model behavior would be to create time series graphs showing the values of each selected diagnostic variable (response, exploratory) over time at 5 year intervals.

Calibration Steps

Identify parameters and associated diagnostic variables

In this step, a comprehensive list of parameters is compiled to create a framework for developing a calibration strategy (see next step). In addition, a set of diagnostic variables must be identified for all of the parameters to use as the "gauge" in adjusting each parameter. The specific parameters and their associated diagnostic variables should be selected based on criteria developed by the modeling team, such as: (1) their importance in understanding response and explanatory variable behavior; (2) their importance in their impact on the key processes simulated in the model; (3) their importance in successfully completing project objectives; and most importantly, (4) the amount of reference information available to compare and modify the parameters in question. The most important aspect about variable evaluation is that the user and modeling team should have a general knowledge of the nature of the variables because they can't evaluate realism if they have no idea how the variable acts in the real world. Moreover, the variables selected to be diagnostic should also be well represented in existing data, literature, or textbooks. The diagnostic variables associated with each of the final parameters should be clearly documented. A common way to document and organize the parameters and associated diagnostic is to use a spreadsheet which can subsequently be used to track the iterative calibration process (next three steps) (Table 6.1).

Let's use the example above for illustrating this step. Since the modeling team has identified basal area as the response variable, it was decided to concentrate calibration on parameters in the growth algorithm, such maximum diameter at breast height (DBH_{max}) and height (HT_{max}). For response variables, they decided that the proportion of landscape area dominated by each species was appropriate (Figure 6.2, 6.3). In addition,

Table 6.1. An example of a spreadsheet of parameters and associated response variables to document the calibration process. DBHmax-maximum diameter at breast height (cm); Agemax-maximum attainable age (yr); Q10-the Q10 coefficient in respiration function; NC-No change; AP-acceptable parameter; OP-overpredict; UP-underpredict.

Algorithm	Parameter	Species	Value	Adjusted	Result	Final
Growth	DBHmax	Douglas-fir	310	320	NC	N
			320	350	AC	Y
		Ponderosa Pine	410	500	OP	N
			500	450	OP	N
			450	425	AC	Y
Mortality	Agemax	Douglas-fir	500	540	NC	N
			540	560	AC	Y
		Ponderosa Pine	610	700	OP	N
			700	650	OP	N
			650	625	AC	Y
Respiration	Q10	Douglas-fir	2.0	2.4	NC	N
			2.4	2.2	AC	Y
		Ponderosa Pine	1.9	2.5	OP	N
			2.5	2.3	OP	N
			2.3	2.2	AC	Y

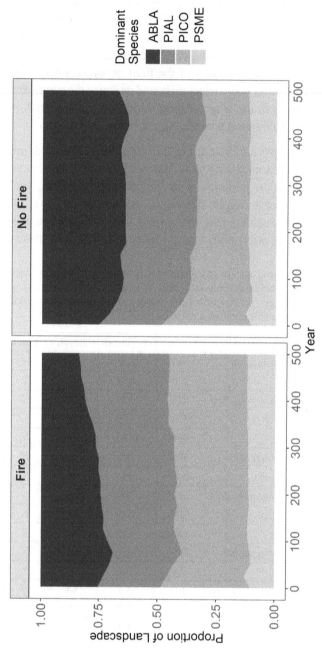

Figure 6.2. An example of a figure often used in calibrating vegetation models. This is a time series graph of the various levels of abundance of each tree species on the landscape coded by shade and shown for two scenarios—fire and no fire. The abbreviations are as follows: PSME-Douglas-fir, PIPO-ponderosa pine, POTR-quaking aspen, SHRB-all shrubs, PICO-lodgepole pine, ABLA-subalpine fir, JUCO-juniper, PIAL-whitebark pine. The abundance measure is the proportion of the landscape occupied by each species determined by the plurality of basal area, except for shrubs which are only possible if there is less than 1 m^2 ha^{-1} of basal area for all other tree species.

they included basal area mortality by species to further understand growth dynamics. The fire suppression specialist also wanted to include parameters related to the fire regime, which includes fire return interval (FRI) and mean fire size (MFS), and evaluated their realism using the response variables of annual area burned and basal area lost from fire by species. And because fire regimes may be related to fuel conditions on the simulated landscape, which might impact FRI, the team also added fine and coarse woody fuel loading (Figure 6.4), area burned by fire, and canopy bulk density (Figure 6.4) The primary user, an ecologist, had studied successional and demographic dynamics of forested ecosystems and knew how forested landscapes change over time and decided to also include basal area by species in trees under 1 cm DBH to represent regeneration as additional diagnostic. The parameter and response variable selections were entered into a spreadsheet along with the values of all parameters (Table 6.1).

Develop calibration strategy

It is important that a comprehensive calibration strategy be developed to minimize subjectivity, increase model performance, gain deep knowledge of model behavior, and obtain information for interpretation of the modeling project's results (Gupta et al. 1998). The strategy can be simple and target only one diagnostic variable for evaluation, or it can be complex and involve an intricate simulation design to set the stage for the evaluation of multiple diagnostic variables.

A well-designed calibration strategy involves creating a simulation plan specifically designed to prod the model to generate results for contrasting investigative scenarios to compare against existing data, findings in the literature, and the experiences and knowledge of the modeling team. There are areas in which the user and modeling team will have great wisdom and the calibration should exploit these areas to ensure realistic results are being simulated. This will provide additional confidence, to both the user and the team, that the model is acting properly. Once the strategy is formulated, it is important that it be documented and followed closely so that subjectivity is mitigated.

The first step is to come up with a set of contrasting simulation scenarios that capture comparative model behaviors in a context that is well understood by the modeling team. For example, two juxtaposing scenarios with and without fire simulated for 200 years might allow the modeling team to contrast successional dynamics and the effect of fire on the landscape (Figure 6.2). These contrasting scenarios could be designed to exploit the range of simulation results for a specific parameterization to positively identify any parameters that might be inaccurate. Examples might be contrasting climates (warm vs cold) (Figure 6.3), management strategies (prescribed burning vs thinning), and initial conditions (bare ground vs contemporary conditions) to evaluate changes in species productivity.

The second step is to decide how long to run the model to capture parameter behavior. This, of course, depends on the data, research findings, and the collective knowledge of the project team. If data spanning a decade is available, then the model should be executed for that length of time. However, if all that is available are long term trends and coarse evaluations of disturbance regimes, then the model should be evaluated over hundreds of years. The user must match the simulation time with the temporal depth of the collective available knowledge.

A last important feature of a well-crafted calibration strategy is the format of the evaluation materials. How will the results from the model runs be synthesized and displayed for focused evaluation? Here, possible options are figures, graphs, tables, maps and any other format that may help with evaluation. Time series graphs were used in the majority of the simulation projects in which I was involved (Figures 6.2, 6.3, 6.4), but several other options may be more insightful for modeling projects. For example, a response variable may be plotted against all of the other diagnostic variables using scattergraphs using each year as an observation. Alternatively, a table may be created that presents a statistical summary (mean, range, standard error) of various diagnostic variables by different zones on the simulation landscape and by different plant functional types. Maps of the values of diagnostic variables can also be created for five

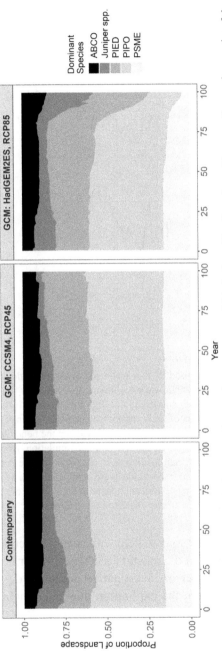

Figure 6.3. Another example of using vegetation as a key response variable in the calibration phase. These are time series graphs of the proportion of the landscape dominated by various tree species by simulation year for three climate scenarios—historical, a IPCC (2007) B2 climate (hot, wet) and A2 (hot dry). Species dominance was estimate by the plurality of basal area.

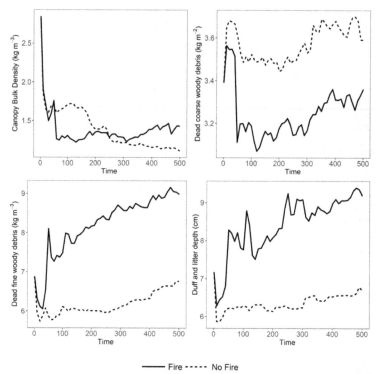

Figure 6.4. An example of the type of output product that was used to evaluate model results for model calibration. A truncated set of diagnostic variables are graphed over time for two contrasting scenarios (no fire=dashed line) and fire (solid line) to determine if the model is acting properly or realistically. Here a set of fuels variables (canopy bulk density, coarse woody debris, fine woody debris, and duff+litter) are used to evaluate the behavior of the model under a non-fire and historical fire regime over 500 years of simulation. Note that the duff+litter is represented by depth rather than loading as in the other surface fuel components (coarse and fine woody). This is because the results were being evaluated by a fire specialist and silviculturalist and those units were understandable to both. Often there will be many graphs for all the diagnostic variables selected for evaluation.

time intervals during the simulation time span to help with spatial patterns. Users should keep in mind the number of diagnostic variables, the number of calibration scenarios, and the number of investigative products (e.g., graphs, tables) that will be created from the calibration may overwhelm the modeling team.

Collect available information

The next step in calibration involves compiling all existing information on the target simulation landscape and ecosystems contained within so that model results can be compared to some semblance of realism (Janssen and Heuberger 1995). Available information might include existing data, findings in the literature, and the experiences and knowledge of the modeling team. This compilation need not be as extensive or as demanding as it was for the two preliminary phases (initialization, parameterization) and the other intermediate phases (validation). Usually it involves compiling a set of reference material from the literature to employ when interpreting model results. Papers related to the model, the modeling project, and the simulation landscape should be assembled, especially those from local ecological studies. Hopefully, information gathered from the previous two phases (initialization and parameterization) can be used again. As always, special attention was should be given to the fields of expertise of the modeling team. Experts should be identified and then warned that they may called upon to evaluate model output if the primary user does not feel comfortable in interpreting results. All reference material should be documented and be readily available for the modeling team.

Carrying on with our example, there was ample information in the literature and from local studies on the historical vegetation and disturbance dynamics of the landscape. The ecologist user identified all possible documents or data sets that can be used as reference in evaluating calibration runs, such as published findings, science textbooks, local reports, environmental impact statements, and related materials. Published papers were found in the fields of fire, hydrology, wildlife, and fuels that matched the expertise of the modeling team, and several of the papers were from landscapes near the simulation landscape. Especially important were a set of publications that contained time series of the response variable and two exploratory variables. Copies of these documents were stored on the computer under an understandable directory structure and the citations for these documents were included in the citation software for later use in the project. Keywords were added to the citations in the software for ease of searching. The team was now ready to start performing preliminary simulations.

125

Perform preliminary simulations

The next step is to see if the model is running correctly and producing realistic results, and this step sets the stage for selecting those parameters that might be adjusted to improve model behavior. Here, the user would run the model under the devised calibration strategy and examine selected variables (i.e., diagnostic variables). Simulation results are then synthesized into the desired output products specified in the calibration plan and then the results are carefully evaluated for plausibility. Realism is determined from the information collected in the previous step and, more importantly, from the modeling team and primary user's experience and knowledge. If unacceptable results are identified, the user and team would then identify the parameters that most influenced these results (see next step). A list is created of those questionable parameters that might be adjusted to make results more credible.

Continuing with our previous example, the ecologist user has executed the model for all four of the calibration scenarios using the suite of diagnostic variables identified by the team. Model output was synthesized into a set of time series graphs and the graphs were imported into a word processor and that file was distributed to the modeling team to be evaluated for realism. Team members then discussed the results that they thought were unrealistic and they identified those parameters that might be the culprit. All parameters in the growth algorithm, such as DBH_{max} and HT_{max}, were listed along with several soil and fire parameters. The team populated a spreadsheet with these parameters (Table 6.1).

Compile a final list of parameters and associated diagnostic variables

In this step, the modeling team must decide on a final set of parameters and corresponding response variables to complete calibration; the evaluation spreadsheet (Table 6.1) is reduced to produce a final set of parameters to use in calibration. The user and modeling team may decide to remove some parameters from the original list either because they are correlated to other parameters, or there simply is not enough time to evaluate every

126

parameter. There is never enough time to calibrate every parameter in a complex ecological model, so the final list of parameters should account for the allotted time for calibration, the expertise of the modeling team, the available computing resources, and the amount of reference materials.

Let's use the example for illustrating this step. The modeling team thought that the simulated time series of basal area created in the previous step did not make sense for the second scenario (no disturbances)—the basal area had a downward trend over time. The downward trend was also present in the first scenario (historical fire) but it was more pronounced. Therefore, the team decided that the two growth parameters should be in the final list (DBH$_{max}$ and HT$_{max}$). The fire return interval (FRI) parameter was also selected because the team thought the landscape may be burning too much thereby causing the basal area decline; mean fire size did not seem to be important in fire dynamics. And because a decline in basal area might cause unrealistic fuel conditions on the simulated landscape, the user also added fuel loading to the final list.

Adjust parameters

The strategy has been finalized, the calibration scenarios have been designed; all reference material has been gathered; the preliminary runs have identified those parameters that may be inexactly quantified; and diagnostic variables for those problematic parameters have been identified. Now it's time to actually adjust a parameter, run the model (next step), evaluate the results (step after next), and repeat these three steps until the model is generating acceptable results for that parameter. This iterative process starts with the first parameter in the parameter list, and then is repeated, one-at-a-time, for each of the parameters that have been identified for adjustment—a series of iterative loops. The user would then go through the entire parameter list one-by-one, calibrate each parameter using their specialized diagnostic variables as computed across each scenario, and evaluate the resultant output.

The question asked by many users is "how much do I adjust the parameter in question?" This iterative process is often a trail-and-error method and involves subjective guesses as to the magnitude of increase or decrease in the model parameter to generate more believable results (Vanclay and Skovsgaard 1997). The user should treat this as a process rather than a one-time adjustment. First, the user has to decide which direction to change the parameter value (up or down), and this is based on: (1) experience with this algorithm, (2) knowledge gleaned from the literature, (3) estimated model uncertainty, and (4) consultation with other modelers or users (Gupta et al. 1998). Once the direction of adjustment is identified, then the magnitude of that adjustment must be determined. If the preliminary results are deemed only a "little" off, then a small adjustment may be indicated, but if there is little knowledge of the behavior of this parameter's algorithm, then a more systematic and less subjective method might be warranted. Some modeling projects use a "half-double" approach where the parameter is decreased by a half or increased by double, and this is repeated iteratively until an acceptable value is achieved. This may seem extreme and time-consuming, but remember, calibration is an iterative process and the valuable knowledge learned from each iteration provides background and confidence for the completion of later phases. Other adjustment approaches can also be implemented, such as a percent increase/decrease ($\pm$ 20%), a proportional increase/decrease (ratio of the parameter to response variable), or subjective guess. More complex adjustment procedures are also available (Beven and Binley 1992). Users should also be aware that responses to adjustments may not be linear.

In our example, the user takes a half-double adjustment strategy and doubled the DBHmax of the species present on the simulated landscape upwards by 2.0 multiplier. Then the user made an educated guess that the FRI for the simulated landscape was too low and deduced that this frequent fire did not allow tree regeneration to become mature trees because the trees were burned before they reached a survivable size. The original value (FRI=25 years) was increased by 50% to FRI=38. These adjusted values were entered into the project spreadsheet.

Run model

In this step, the model is simply run with the newly adjusted parameter. Three things are important in this step. First, each parameter value, before and after adjustment, should be documented to keep track of the adjustments during this very complex iterative procedure (Table 6.1). This can be done by saving each input file in a specialized directory structure, recording the adjusted parameter in a spreadsheet, writing the value in the simulation notebook, or typing the value into a word processing document. Many calibration exercises were delayed because the same steps were repeated because an adjustment record was unavailable. Next, the output, and associated graphs and figures, should be stored in a directory structure or system that identifies it with the parameter adjustment. A spreadsheet such as Table 6.1, could identify the new parameter value and also the name of the file where the results of the simulation with the new parameter are stored. And last, details of the model run should be noted, including the time it took to simulate the scenario, the amount of memory it took, and the amount of output generated. These could be important in the execution phase.

To continue with our illustrative example, the user ran the model for 50 years for the four scenarios in the calibration strategy. Values for eight diagnostic variables were reported for each year over the 50 years which created a large data set (8 variables X 50 years X 4 scenarios=1600 values). Output data were stored in a directory labeled "Interation1" to keep track of the results during the iterative modifications. A spreadsheet was updated with the newest parameter value and a link was included in the spreadsheet to the resultant reports from the output.

Evaluate results

This step involves the inspection of the new results after parameter adjustment. Some people on the modeling team won't feel comfortable assessing model results, in which case, they should consult other local experts about the realism of the results. An outside person that should be consulted during the calibration process is the developer of the model. If that person is unavailable, then other users of the model should be consulted. Again, the graphs and tables are the "gauges" on our "dashboard" to evaluate model performance. If the parameter is still deemed to be in error, the previous two steps are repeated until the parameter value is acceptable.

And once that parameter is calibrated, the next parameter in the list is calibrated by iteratively repeating these last three steps.

In our example, after adjusting the DBH_{max} parameter and running the model, the user created the basal area, DBH, and fuels time series graphs (see examples in Figures 6.2, 6.3, 6.4) and passes them out to the team. This process created over 30 graphs (8 variables X 4 scenarios). After evaluation, the new DBH_{max} parameter seemed about right based on the team's evaluation, but the FRI value seemed high as too many trees were becoming established on the site, so FRI was assigned a new value of 31 (a mid-point between the previous two variables) to evaluate if that improved performance. Another model run revealed that the FRI was just right, and now the modeling team went to the next parameter, and then the next, and so on.

Calibration Issues

It is vitally important that the user document each step of the iterative calibration process. A project journal should be started and notes about model behavior and parameter adjustments should be recorded as needed. We often created graphics files (e.g., jpg or png) of model output and stored them in nested file directories named for specifics of the iterative step. These directories or files should be labeled as to version and iterative step in the calibration process. For example, a file whose parameters were adjusted to increase drought tolerance under colder climate conditions might receive the name parm_dry_cold.txt. All journal notes, adjusted data files, and graphed output must be stored using a heuristic organization so that steps aren't repeated and adjustments can be quickly made. As mentioned throughout, spreadsheets are indispensable to organize the calibration details (Table 6.1 for example).

How do you know when you are done with calibration? Many calibration efforts linger on because of the overwhelming number of evaluation products created by the numerous diagnostic variables, the confusing and conflicting calibration output (adjusting of one parameter may improve the behavior of one diagnostic variable but worsen another variables behavior), and the inability to decide if the

130

results are truly realistic. This is again when science suddenly becomes art, and the modeling team must use subjective judgements to decide when a parameter is sufficiently calibrated. Some modeling teams set "drop-dead" date for calibration while other teams vote to determine when calibration is done. In my opinion, a calibration is never done, especially for highly complex, mechanistic models, so it is better to set goals and guidelines for termination rather than to calibrate until satisfied.

On a similar note, users should be wary of tangential investigative avenues that pop up during the calibration process. It may be difficult to stay on task while calibrating the model because new model behaviors often lead the team down interesting pathways—finding out why a parameter is wrong often adds additional time on an already tight project schedule. However, if there is time, users are encouraged to follow their curiosity and gain additional insight into model behaviors. A last warning is that users should try to not be discouraged during a laborious calibration process and focus on the importance that the calibration adds to the project rather than the subjectivity in the calibration. Sometimes it may seem like calibration is trying to adjust a blob of jello, yet many users are often appreciative of the results at the end of the calibration effort.

It is also common that the same analyses used in the validation and analysis phases are used during this calibration phase. Some of the same graphs, figures, tables, and maps may be common across all three phases, and because of this, figures from calibration are often used in the project's final report. Therefore, it is important that all figures are of publishable quality. And often there isn't time to examine all of the materials used in the calibration phase, so the final analysis should concentrate on a set of response and exploratory variables that are important to the project and are easily changed by modifying a parameter, or a set of parameters. Algorithms with a large number of parameters in their design will be extremely hard to calibrate in this phase.

References

Beven, K. and A. Binley. 1992. The future of distributed models: Model calibration and uncertainty prediction. Hydrological Processes 6: 279–298.

Gupta, H. V., S. Sorooshian and P. O. Yapo. 1998. Toward improved calibration of hydrologic models: Multiple and noncommensurable measures of information. Water Resources Research 34: 751–763.

IPCC. 2007. Climate Change 2007—the Physical Science Basis vol Working Group I. Climate Change 2007 Working group I contribution to the fourth assessment report of the IPCC. Cambridge University Press, New York, New York, USA.

Jakeman, A. J., R. A. Letcher and J. P. Norton. 2006. Ten iterative steps in development and evaluation of environmental models. Environmental Modelling & Software 21: 602–614 doi:https://doi.org/10.1016/j.envsoft.2006.01.004.

Janssen, P. H. M. and P. S. C. Heuberger. 1995. Calibration of process-oriented models. Ecological Modelling 83: 55–66.

Rykiel, E. J., Jr. 1996. Testing ecological models: The meaning of validation. Ecological Modeling 90: 229–244.

Vanclay, J. K. and J. P. Skovsgaard. 1997. Evaluating forest growth models. Ecological Modelling 98: 1–12.

7

Validation

Determining Model Uncertainty

"We demand rigidly defined areas of doubt and uncertainty!"

Douglas Adams, The Hitchhiker's Guide to the Galaxy

Validation—the action of checking or proving the validity or accuracy of something (OED).

ABSTRACT

Validation is the task that quantifies uncertainty in model results. Validation differs from calibration in that no parameters are adjusted. Model validation is needed for several reasons: (1) provides the user a valuable estimate of model uncertainty, (2) gives the modeling team, organization leadership, and the general public a measure of confidence, or skepticism, in the simulation results, (3) identifies bugs, problems, or limitations in the model that were previously unknown, and (4) illustrates what organizational data is needed for inventorying and monitoring the response variables in the future. The chapter presents a set of five levels of detail in which to conduct a validation and then enumerates the four steps of validation: (1) collect data, (2) format data, (3) run model, and (4) perform analyses. A set of analysis procedures are provided. Another form of validation is a sensitivity analysis, and details on conducting and interpreting a sensitivity analysis are also presented. Last, validation concerns are addressed.

Introduction

Validation is the task that quantifies uncertainty in model results so that users can integrate this knowledge into the interpretation of those results for completing project objectives (Rykiel 1996, Gardner and Urban 2003, Jakeman et al. 2006). This involves using the calibrated model to generate results that are then statistically compared to reference conditions (field-measured ecological characteristics) to determine a level of accuracy or precision in model outcomes. This task usually requires a dataset that was NOT employed in the initialization, parameterization, or calibration procedure. Validation is different from calibration in that different datasets are used and the model is NOT modified based on the validation results. Most users will experience an overwhelming desire to modify model parameters to improve results during validation (Chapter 6), but, if the model and parameters are modified, then any estimates of uncertainty are gone and the validation has to be redone with another dataset. It is important that a model be validated for each application or modeling effort because parameter sets are different for each project. Previous validations of a model are important as reference, but they are not always applicable when a different parameter set is used.

Model validation is needed for several reasons. First, it provides the user a valuable estimate of model uncertainty, which is critical when interpreting the results. Next, it provides the modeling team, organization leadership, and the general public a measure of confidence, or skepticism, in the simulation results. A comprehensive validation effort can also provide additional experience to the user and modeling team in using and understanding the model. It also can identify bugs, problems, or limitations in the model that were previously unknown. And last, a good validation will identify organizational data needs for inventorying and monitoring the response variables in the future.

134

Use of the term "validation" has been extensively debated in the modeling literature. Some modelers feel the word "evaluation" or "testing" should be used instead of validation because a model's behavior is never really "validated" (Augusiak et al. 2014). Some feel that the root of validation—valid—implies legitimacy (Oreskes and Belitz 2001) and therefore validation implies that the model is deemed acceptable for its intended use because it meets specified performance requirements. Augusiak et al. (2014) believes that evaluation should be fused with validation to make the process easier to understand among decision makers. Verification is a term often used in the modeling literature to describe the process of assessing model accuracy (Oreskes et al. 1994). Rykiel (1996) believes that this debate about the term "validation "arises as much from semantic and philosophical considerations as from the selection of validation procedures". He further states that "validation is not a procedure for testing scientific theory or for certifying the 'truth' of current scientific understanding". All of these are excellent points, but validation is the term selected to be used in this book because its definition "the action of checking or proving the validity or accuracy of something (OED)" seems to fit the topic of this chapter. And while Rykiel (1996) believes validation is an optional task, I feel it should be required on all modeling projects that are used for management decisions even though I've often been remiss on performing validations because of time and data constraints.

Some terminology must be introduced to understand the material in this chapter. First, most validations involve the comparison of observed vs predicted variables. **Observed** values are reference data actually measured in the field or quantified by some other means (see next section). Remote sensing fields call these "ground truth" and other ecological studies call them "reference data". **Predicted** values are the output generated from the model.

As much as we want it to be true, there are never enough data of sufficient resolution, quality, and compatibility to properly validate ecosystem models, especially LESMs and other spatial models. As Oreskes et al. (1994) states "verification and validation of numerical models of natural systems is impossible". This is because ecosystems and landscapes are never closed and because model results are rarely unique. He further states that "models can be confirmed by the demonstration of agreement between observation and prediction, but confirmation is inherently partial". I've found model validation efforts are often limited because of one or more of the following:

1. **Temporal depth.** Often, data are available for only a limited time span, a few years or decades, but the model is being used to simulate for centuries or millennia. The weather for a short time span, for example, may fail to represent the climate over a century of simulation.

2. **Spatial domain.** Data are available for only a small part of the project area or for a limited set of conditions being simulated. For example, data exist for ponderosa pine types but not for Douglas-fir types, both of which are simulated in the model.

3. **Scale and detail inconsistencies.** The scale and resolution of the reference data may not match the scale and resolution of the simulated variables being validated. Empirical data, for example, were described in ways that are incompatible with simulated values (e.g., basal area was only measured for trees greater than 10 cm DBH).

4. **Specialized model variables.** The model may generate values for variables that are difficult to measure in the field, such as net primary productivity (NPP), potential evapotranspiration, and root growth. Moreover, the model's output may include elements not found in the empirical data; simulated dead carbon, for example, must be compared with wildland biomass loadings but measured biomass pools may not be the same as pools for dead carbon.

5. **Inappropriate measurement.** The measurement of the reference data may be inconsistent with model output. Model output may be compared against data that were outside the range of the original development data represented in the model. For example, a photosynthesis equation was only developed for temperature ranges of 35 to 90°F, yet input data has days below freezing and above 100°F. Or, measured respiration estimates include both autotrophic and heterotrophic sources, yet only autotrophic estimates are only output from the model.

Validation is a critical task in a modeling project, but this phase is often ignored by many users and modelers because the above data limitations often overwhelm many validation efforts. I believe that any evaluation of model performance, no matter how limited, trivial, or narrow, will be of value when interpreting model results. Therefore, several approaches to validation are presented to supply the user with enough ideas to conduct beneficial validations to evaluate model uncertainty.

Validation Approaches

Because of the aforementioned data limitations, a possible scheme that users can employ to perform a useful validation is proposed. There are basically five levels of rigor in conducting a validation (Table 7.2). The melding of all five approaches is often needed to comprehensively evaluate model uncertainty. Any approach, no matter how restricted, is better than none at all.

Table 7.1. A summary of the various levels of rigor for conducting model validation. The levels are listed in order of preference. For the example, the context is as follows: a complex mechanistic model is used to simulate changes in tree basal area and volume over 50 years.

Level	Name	Description	Example
I	Primary	Direct comparison of simulated vs observed response variables	Simulated basal area is directly compared with observed basal area
II	Association	Indirect comparison of simulated vs observed response variables	Tree ring growth data is summarized to estimate basal area to compare observed with predicted
III	Module	Comparison of intermediate output from a model's module or algorithm with observed values	Intermediate output from the photosynthesis module is compared with measured photosynthetic rates for trees in the stand
IV	Sensitivity Analysis	Simulation of model over systematic changes in selected parameters to evaluate changes in response variables	Parameters in the photosynthesis, respiration, and water use modules are incrementally changed and subsequent results are compared
V	Opinion	Experts evaluate modeled output	A panel of experts evaluate simulation results to estimate levels of uncertainty or reliability

Table 7.2. An example of a contingency table taken from Lutes et al. (2009b). Note that each of the rows and columns are categorical classes of fuel models.

		Predicted										Total plots	Number misclassified	Percent misclassified
		1	2	3	4	5	6	7	8	9	10			
Observed	1	1193	100	32			1					1326	133	10.0
	2	34	357	24		5	7	4				431	74	17.2
	3	60	2	377	29	113		24				605	228	37.7
	4	4			32							36	4	11.1
	5	22	31	95		91	12	20	1			272	181	66.5
	6	10	4			13	238	21	15	2		303	65	21.5
	7		3	20		76	44	370	83		4	600	230	38.3
	8	1					34	12	187	47	2	283	96	33.9
	9	5					2	2	23	132	10	174	42	24.1
	10									1	15	16	1	6.3
	Total											4046	1054	26.1

Level I

The best approach to validating a model is to compare the simulated response variables identified in the modeling objective (Chapter 3) to actual observations or measurements of the same variables within the area in question over the time period represented in the empirical data (Table 7.1). This is the Holy Grail of all validation efforts. When simple models are used, such as S&T or growth and yield models, there may be sufficient data available to accomplish an acceptable validation. But, more often, one-to-one comparisons of simulated to empirical response variables are difficult because of the data limitations. Therefore, modelers have developed some alternative approaches.

Level II

The next level involves comparing response or explanatory variables with surrogate empirical data (Table 7.1). Here, a direct correspondence between predicted and observed variables might be missing for response variables, but there could be other available data sets that may be associated with the simulated variables to provide a useful comparison. In a simplistic example, the cover type categories that are simulated by an S&T model may be different from the categories available historical maps, so a cross-walk table could be developed based on field data to link the simulated cover type categories to the empirically mapped categories. Some simulated cover types may not have a common category across the predicted and observed so they can be removed from the analysis. Another variation of a Level II approach is to collapse scales to mesh empirical data with predicted data. Here, the desired reference data may be available at annual intervals or for a few stands or for small areas. Therefore, the user can summarize predicted variables to match the scale and detail of the reference data. Annual estimates of burned area, for example, may be unavailable for the simulation landscape, but fire history studies can be used to approximate average area burned per year from fire-scarred trees (Liu et al. 2011) then model results are synthesized to correspond to with results from fire history studies. In another example, I was involved in a project where we used MODIS NPP layers to validate modeled NPP over a high elevation landscape using FireBGCv2 only to find that the variation of predicted NPP within the MODIS 500 m square pixel overwhelmed the comparison.

Level III

Another approach is to validate output from a critical module or algorithm in the model rather than validate higher order variables, such as response variables (Lorscheid and Meyer 2016). Data may only be available to compare a small element of a model's simulation, such as predicted leaf area, snow depth, or monthly evapotranspiration, so the user outputs those intermediate variables from the simulation and compares them to the available empirical data. These are not response or explanatory variables, but instead they are intermediate results from a particular simulation, and they may have nothing to do with the objective. The assumption here is that if one or more of the model's algorithms are acceptable, then there may be higher confidence in the response variable output. In a couple of examples, Keane et al. (1996) validated simulated tree ring growth with actual measured tree chronologies in one effort and validated fire regimes from fire history studies. Holsinger et al. (2014) used streamflow measured by USGS to compare against simulated watershed outflow in the FireBGCv2 model to evaluate accuracy for a fish habitat study. The user should attempt to find data for the most critical algorithms first, and this can be identified from a sensitivity analysis (see next sections) or from the literature, and if this is unavailable, users should try to find data any possible algorithm, even less important modules.

Level IV

Another approach to validating models is to conduct a **sensitivity analysis.** A well-constructed sensitivity analysis can yield interesting insights into the behavior of the model and provide background into interpreting model results. While this approach does not directly assess accuracies or precisions of model predictions, it will help in establishing a level of confidence in model predictions. A general description of sensitivity analyses procedures is presented in a later section in this chapter.

Level V

The last approach is the least accurate but perhaps the most popular. This involves having one or more experts evaluate model output to estimate a level of reliability and uncertainty in the predictions (Table 2.1). Here, a series of graphs and tables are created, similar to ones used in the calibration (Figure 6.3) or analysis (Figure 9.5) phases, and then given to the "validation" team for evaluation. They can rate the reliability, believability, or uncertainty of simulation results using a scale similar to the one presented in Table 2.1. This is a last resort approach and should be done only if few data can be found to help evaluate the model. This type of evaluation is somewhat contrary to the very reason a modeling project was needed; if the "experts" know the answer, then why do the modeling project? However, it is always best to have broader reviews of the modeling results to aid in output interpretation, and the expert team can provide excellent input into model behavior, and also in other areas of the modeling project as well, such as the length of simulation, the selection of response variables and their units, and possible avenues of data exploration (Chapter 9). Variations of this "Delphi" approach (Orsi et al. 2011) include literature syntheses (i.e., using information from the literature to compare with model findings), comparisons with other modeling efforts (i.e., using results from a previous validation of the model), and comparisons with results of a similar model (i.e., using validations of similar models).

Validation Steps

Unlike calibration, validation should be done with a sound dataset that hopefully has been compiled from actual data measurements in the field. The following is a step by step approach on conducting a validation. The steps involved in collecting the data are nearly identical to those presented in the initialization (Chapter 4), parameterization (Chapter 5) and calibration (Chapter 6) phases but the criteria for selection is somewhat more focused. These steps may be slightly different depending on the level of validation discussed above.

Collect data

In Level I validations, field data provide a wonderful reference for simulated data and this is the best validation approach. Here, data are collected for the simulation area to compare against simulated predictions, especially the response variables. Tree chronologies, for example, can be constructed from tree ring cores and the radial or basal area growth can be compared against model predictions. Carbon dioxide and water flux measurements from eddy flux towers can be compared with predictions from mechanistic ecosystem process models (Amthor et al. 2001).

If there is time and funding to conduct a field campaign for collecting validation data, then here is the summary of the important suggestions from Chapter 3 that might make that effort more effective:

1. Merge data collection with the three previously mentioned phases (initialization, parameterization, and calibration).
2. Ensure field data are useful to your organization.
3. Collect data that may be used to understand validation results.
4. Use methods and protocols that mesh with organizational requirements.
5. Confirm that the sampled data can be used for validation.

If there are no corporate data concerns for your organization, there are a number of standardized methodologies could be used as templates for field sampling in validation field campaigns, such as FIREMON (Lutes et al. 2006) and FFI (Lutes et al. 2009a). It is useful if the selected sampling system include sampling designs, protocols, field codes, databases, and possible analysis reports. These entered field data should be easily exported from the package into the statistical packages that are used to conduct validation analyses.

If collecting data is impossible, then there should be a comprehensive search for existing data, and this search should be extensive but flexible (Level II). The temporal and spatial domain could be modified to accommodate possible data sets that could be used in the validation. High priority data sets, for example, would come from the area bounded by the simulation landscape, but if data exist from outside or near the simulation landscape, the model parameter set and initialization data could be modified to

142

approximate the new spatial domain. Similarly, simulation specifics of the model can be modified to reflect the temporal scale and domain (reporting time, span of time) of a possible validation data set. A data set may have the wrong units or a coarser spatial scale than modeled output, so additional formatting of the data set can be done to more closely match the predicted vs observed formats. There may be many sources of data that could be used for validation and users should first evaluate possible data sets that have been compiled by their organization or agency. These will probably be the easiest and fastest to attain. Then, universities, other government agencies, and private organizations (NGOs) can be contacted. A literature review may also be especially useful as it can be used in all other modeling phases, and it may point to various datasets collected for research or management purposes.

Measured field observations are the most desirable validation data, but validations can be done with other data types (Level II). Remotely sensed data could be used for both Level I and Level II evaluations. Ecosystem productivity estimates such as NPP, GPP, and LAI (leaf area index), are remotely sensed from a variety of MODIS, AVHRR, and Landsat platforms. Information in the literature can also be used to calculate a model response or exploratory variable from remotely sensed imagery. However, nothing takes the place of field measured data. These data can be found from previous ecological inventories, such as timber assessments, population demographic surveys, species abundance studies, or they can be sampled as part of the modeling project.

Users will often find that few data sets exist to validate their modeling project, so another level approach might be warranted (Level III–V). Findings from studies in the literature can be compared with model results or a panel of experts can be convened to evaluate model results, but this evaluation should be carefully planned and the specifics of the simulation design should be specially tailored to the evaluation experts (Level V). If there are data sets, but they are somewhat limited, such as in variable representation, temporal depth, and spatial coverage, then it is suggested that the user combine several validation approaches into the validation effort. If data exists for only one response variable, for example, but there are ample data available for module validation (Level III), then a series of module output can be validated along with the single response variable (Level I), but in addition, an expert panel can evaluate the remaining response variables (Level V). If funds and time exist, then a sensitivity analysis may be conducted to augment the validation (Level IV).

Format data

Once all possible data sets have been identified, it is time to reformat the data for direct comparison to model output and to reformat the model's initial conditions, parameters, and simulation specifications to reflect the domain of the validation data sets. The reformatting of existing data are discussed extensively in other chapters and can be done with existing software packages. Care should be taken to ensure the right units, scale, and domain is reflected in the reformatting of the data.

Adjusting initial conditions to reflect the characteristics of the compiled validation data is one way to ensure that the reference data will correspond directly to the simulated data. For example, the US Forest Service has a comprehensive Forest Inventory and Analysis (FIA) program where sampling sites are established on a 5 km grid across the US and trees on these sites are measured at staggered time intervals (Bechtold and Patterson 2005). If validation involves comparing measured tree or stand growth to simulated tree or stand growth, then the model has to be initialized for the date at which the trees were first measured and the model should be executed until the date of the second measurement. Moreover, the weather record should reflect that time span. The model should then be executed to simulate tree and stand growth for the same years over the re-measurement period. For example, we often created stand conditions (tree size and density) in the past by growing the trees backward using a growth and yield model through a trial and error method (e.g., guess at the diameter 30 years ago and run the G&Y model to see if the diameter is the same as measured today).

Run model

The model can now be executed once all existing data are compiled and formatted. But, to run the model, the simulation parameters must be modified to reflect the limitations and scope of the reference data. The length of simulation, for example, must be adjusted to reflect the temporal domain of the reference data. Moreover, the simulation landscape may need to be divided into smaller units if the reference data is only available for a stand, small area, or a specific plant community. Output variables must be selected to match or approximate the variables represented in the validation data set. One major concern in conducting validations of models with stochastic elements is how many runs should be executed to represent the inherent stochasticity. This depends on many elements of the model and simulation area so an estimate of the number of runs is difficult, but at least five replications is a good starting point. More on validation of stochastic models is presented in the Issues section.

Perform analyses

There are a number of statistical analysis that can be used to validate model predictions depending whether the response variables are continuous or categorical. It is impractical to detail every type of analysis as it is the subject of many statistical papers and books on modeling (Brown and Kulasiri 1996, Zar 1996, Vihinen 2012) . However, the following are some validation analyses that I've found useful in past projects.

Residual analysis

In this analysis, the differences between predicted and observed values are calculated and analyzed using statistical means to provide some insight into model accuracy (Cox and Snell 1968). The mean of the differences of predicted vs observed (P-O) is often called "bias" and it provides a quick and easy estimate of the error in predicted values; perfect predictions have zero bias, while a negative difference indicate model underestimation in the response variable and positive bias indicate overestimation. The absolute value of the differences can be averaged to calculate the mean absolute error (MAE) (Pontius et al. 2008). The magnitude of the MAE can be compared to the mean of the observed values as another quick and easy estimate of error. However, bias and MAE are not especially diagnostic, especially when there are abundant outliers in the validation data set (e.g., a large difference in one observation can dominate an estimate of bias or MAE, especially when there are few observations). Therefore, each difference is often squared and the sum of these squares is analyzed using statistical packages to determine more revealing validation statistics. One statistic that is especially useful is the root mean square error (RMSE), which is the square root of the summed squared differences divided by the number observations. This is perhaps the best estimate of error in model predictions and one that is reported in many modeling papers.

Regression analysis

Another extremely useful tool to compare observed and predicted data is regression analysis (Helmreich 2016). Here, predicted values (dependent variable) are regressed against observed values (independent variable) and results are graphed and summarized using a set of insightful least squared regression statistics (Figure 7.2). The coefficient of determination (R^2;

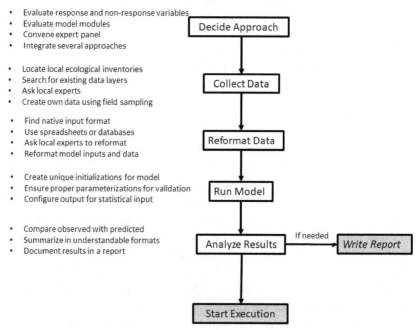

Figure 7.1. The steps used in the validation phase of a modeling project. Steps in italics indicate that the step is optional. Various tasks are summarized to the left of each step.

value between zero and one) is used to determine how well the predicted values are correlated the observed (tight correlations of > 0.80 validate high precision). However, R^2 values do not provide a comprehensive measure of accuracy and should never be used alone for the validation assessment. The real strength in regression analyses is the computation of three other statistics. The slope of the regression line (*beta* coefficient) indicates if the predicted values are overestimating or underestimating the observed; slopes of 1.0 indicate perfect fit, below 1.0 means predictions are underestimated, and overestimation above 1.0 for observed vs predicted. The fitted intercept (*alpha* coefficient) provides an indication of how much bias there is in the predicted values. And last, the variation about the regression line is used as a measure of accuracy and is similar to RMSE when around observed mean. If the variation is high (> 50% of the mean) then the results have low certainty.

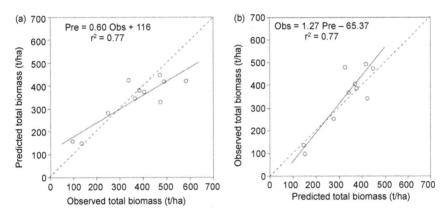

Figure 7.2. An example of using regression analysis to evaluate model accuracy, precision, and uncertainty taken from Piñeiro et al. (2008). The graph show observed vs predicted for a modeling project where the model is over-predicting for values less than 250 t/ha and under-predicting when biomass is greater than 300 t/ha. The slope of the line is 0.6 when it should be close to 1.0, and the intercept is 116 t/ha which should be zero. The R^2 of 0.77 is high indicating high precision, but the accuracy of the predictions is moderately acceptable.

Contingency tables

Contingency analyses are used with categorical data, such as output from S&T models, to determine how often the model predicted the right category (Congalton 1991, Congalton and Green 1999). In this method, a contingency table or confusion matrix is created where predicted classes are the rows and observed classes are the columns (Table 7.2) (Sokolova and Lapalme 2009). The numbers in each cell indicate how many of the modeled predictions for a certain class matched each of the classes of the observed data. The diagonal of the table are the cells where the predicted class matched the observed and the sum of the diagonal cells divided by the total number of observations multiplied by 100 is an approximation of model accuracy—the percent correct. However, a contingency analysis can also provide insight into the distribution of the classes that were wrongly predicted—one class was most often observed when a certain class was predicted. A full discussion of contingency analysis is too lengthy for this book.

One of the most popular metrics used in validation of categorical data is the area under the ROC curve (Fawcett 2006). The ROC curve is created by plotting the true positive rate against the false positive rate at various threshold settings (Figure 7.3). A receiver operating characteristics (ROC) graph is a technique for visualizing, organizing and selecting classifiers based on their performance. ROC graphs have long been used in signal

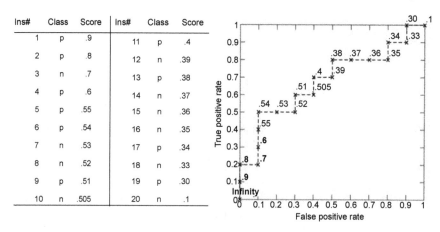

Figure 7.3. The area under the ROC curve provides a diagnostic indication of how well the model is predicting categorical data. Shown are 20 points that are classes where the false positive and true positive rates are plotted. The area under the "curve" provide insight into how well a model is predicting across all classes. This entire discussion of ROC is taken from Fawcett (2006).

detection theory to depict the tradeoff between hit rates and false alarm rates of classifiers. The biggest advantage of using ROC curve is that it is independent of the change in proportion of responders. ROC graphs are two-dimensional graphs in which true positive rate is plotted on the Y axis and false positive rate is plotted on the X axis. An ROC graph depicts relative tradeoffs between benefits (true positives) and costs (false positives).

Sensitivity Analysis

It is best if a sensitivity analysis is done by the person who built the model, not by the user, and it would be nice if it were done just after model development, as well as during a new application of the model (Kleijnen et al. 1992). Unfortunately, these analyses are rarely done by modelers or users because they are just too demanding computationally (Cariboni et al. 2007). Some believe that a sensitivity analysis and comprehensive validation should be included in all journal articles involving ecological modeling (Kleijnen et al. 1992). While sensitivity analysis does not quantify accuracy and precision, it provides a number of other valuable insights into model behavior, such as the importance of critical parameters, the degree of parameter interactions in simulation results, and a level of confidence in parameter quantification (e.g., unknown parameters that we estimated by the user may be insensitive to model behavior).

In a sensitivity analysis, a small change in a major parameter is made and the resultant change in one or more response variables is calculated while keeping all other parameters constant (Saltelli et al. 2008). The same initial conditions are used for all adjustments. The parameter value may be modified across a gradient of possible realistic values (e.g., the plant respiration coefficient is assigned ten values that span the range of values found in the literature), or it can be simply computed as a percent change (e.g., the respiration coefficient is reduced by 10%) (Krieger et al. 1978). A sensitivity coefficient, the ratio of the change in the response variable to the change in input parameter while all other parameters remain constant, is a good diagnostic statistic to evaluate parameter sensitivity (Ambas and Baltas 2012).

149

Approaches to sensitivity analyses have been discussed in the literature (Xu et al. 2004). Here are some sensitivity methods that are commonly used and are summarized based on the structure presented by Hamby (1994):

1. **Single parameter and single response variable.** A single parameter and a corresponding single response variable are used to evaluate sensitivity—often termed a one-to-one comparison. This is a simple, quick, and easy way to conduct a sensitivity analysis on the most important parameter in the model. It is up to the user to pick the parameter and the response variable so there is some degree of subjectivity.

2. **Single parameter and multiple response variables.** Here a single parameter is evaluated over the response of several response variables. This allows the user to evaluate a parameter's influence on several simulated processes to ensure that the model is acting acceptably.

3. **Multiple parameters and multiple response variables.** A set of parameters is selected and each is altered using a systematic approach (i.e., alter one and leave the others constant, reset the first variable and alter another variable, and so on) and the results are compared against one or more response variables. Each parameter-response pair are subjectively selected and evaluated independently of the other pairs.

4. **Factorial design.** All of the previous approaches were one-to-one comparisons and did not evaluate changes of each parameter across all selected parameters and all selected response variables. In a factorial approach, the parameters are considered factors and the levels of parameter change are considered the treatments in statistical terminology. The model is then run for every combination of parameter and parameter adjustment and results are compared across one or more response variables. Because this approach allows the user to detect interactions between parameters, it is by far the most informative and interesting. But, it is also quite time-consuming and the analysis is quite complex.

5. **Response space.** This is an approach that has a factorial design but includes all appropriate parameters in the model and each parameter is given values across the entire range of possibilities using a systematic approach (e.g., percentiles, 10 evenly spaced values between a minimum and a maximum). This should define the range of all possible responses from the model for one or more variables (i.e., response space). This is an incredibly difficult simulation design that requires inordinate amounts of computer time and storage to complete. This is not suggested for the natural resource user.

Many of the methods presented in the previous sections can be used to compare changes response vs parameter changes. Saltelli et al. (2008) suggest that the scatterplots, regression analysis, variograms (exploration of the variance of the change in response variable), and probability distributions are also useful analysis techniques.

A sensitivity analysis is impractical for most modeling projects due to the sheer number of simulations and the high degree of complexity in sensitivity analyses. However, if validation data are scarce, a sensitivity analysis may be the only way to: (1) gain confidence in the parameterization, (2) estimate the degree of uncertainty, and (3) provide experience and understanding into model behavior. However, an insightful sensitivity analysis demands a comprehensive knowledge of the model *a priori* because the user rarely knows which parameters are the most important or critical, and to what response variable are they the most critical. Therefore, it is beneficial if the modeler is available to help with the simulation design of the sensitivity analysis. In fact, the presence of an existing sensitivity analysis might be a good criteria for selecting a model.

Validation Concerns

Perfect validations for any model are impossible and users should accept that any validation will have limitations. Uncertainties in the data available to test models (observed data; measurements) lead directly to uncertainty in the accuracy of model predictions (Amthor et al. 2001). As mentioned for spatial model results, there is a lack of spatially explicit historical time series data that are in the appropriate context to compare with model results. Validation data must have many characteristics to be useful for model validation and most data sets are missing some of these characteristics. As an example, validations of fire spread models are often constrained by accurate measurements of active wind fields at the right temporal and spatial scales because major changes in the spread of wildland fire can be caused by fine scale variations in fuel moisture and wind profiles on surface and canopy fuels (Finney 1998).

Model validation involves the complex melding of several difficult tasks, especially with landscape models that simulate vegetation and fire dynamics over millennial time spans (Keane and Finney 2003, Keane et al. 2015). It may be helpful to present examples of how other modelers have validated their models. Schaefer et al. (2012) used an analysis of residuals to evaluate gross primary productivity measured at various eddy covariance flux towers with a variety of simulation models. Stage et al. (1995) performed an extensive validation of succession parameters for the pathways in the Interior Columbia River Basin Ecosystem Management Project using results of another model (FVS stand growth model) (Crookston and Dixon 2005) and found greater than 80 percent accuracy. Keane et al. (2002) compared simulated fire area and pattern statistics from a 1,000 year LANDSUM run to the historical fire atlas created by Rollins et al. (2001) and found acceptable agreement between the distributions of fire size and patch shape. Amthor et al. (2001) used measured and derived daily CO2, evapotranspiration, and air temperature from a spruce forest in Canada to compare with output from nine ecosystem process models.

Perhaps the most important question in validation is how do you validate a stochastic model or a model with stochastic elements? First, the user should attempt to remove as much stochasticity from the validation runs as possible by crafting validation designs that include the least amount of stochastic processes in model simulations. Wildland fire, for example, is highly stochastic in landscape simulations because of the randomness of fire starts. Therefore, validations should be done without fires or with fire occurring at the same time and on the same piece of land as the observed fires, or validations can be done at longer time intervals to assess regimes rather than events (e.g., fire characteristics over long simulation times can be used to quantify the fire regime, and that regime can be compared with reference fire history data). Using the fire example, the simulated landscape fire rotation, fire return interval, and area burned per year can be compared to studies that evaluated fire regimes using fire-scarred trees (Fall et al. 1997, Parsons et al. 2007). And last, the user can perform multiple validation runs and include these as additional observations in the analysis procedures mentioned in the section above. At any rate, stochastic models must be run several times and each replication should be considered an observation in the validation analysis.

Users should be careful when using and interpreting variance as an absolute measure of uncertainty or accuracy. In highly complex models, especially when results are multi-modal, other statistics may be more appropriate. Smaldino (2013), for example, mentions that entropy, not variance, is better when dealing with discontinuous or complex model output. Moreover, variance contains both model error and natural variation with the contributions of each are relatively unknown. Therefore, it will be difficult to determine if poor validation results are because of highly stochastic modeling algorithms, errors in the model and its parameterization, or high natural variability.

And last, and mentioned in all other chapters, users should document the method, details, and results of the validation effort in a report or in computer files. In past validation efforts, a summary table and figure were often prepared for the modeling team to interpret the final simulation runs. Moreover, a journal will often demand extensive detail in the validation of the model and well-kept records will make this quite easy.

References

Ambas, V. T. and E. Baltas. 2012. Sensitivity analysis of different evapotranspiration methods using a new sensitivity coefficient. Global NEST Journal 14: 335–343.

Amthor, J. S., J. M. Chen, J. S. Clein, S. E. Frolking, M. L. Goulden, R. F. Grant, J. S. Kimball, A. W. King, A. D. McGuire, N. T. Nikolov, C. S. Potter, S. Wang and S. C. Wofsy. 2001. Boreal forest CO2 exchange and evapotranspiration predicted by nine ecosystem process models: Intermodel comparisons and relationships to field measurements. Journal of Geophysical Research: Atmospheres 106: 33623–33648.

Augusiak, J., P. J. Van den Brink and V. Grimm. 2014. Merging validation and evaluation of ecological models to 'evaludation': A review of terminology and a practical approach. Ecological Modelling 280: 117–128.

Bechtold, W. A. and P. L. Patterson. 2005. The enhanced forest inventory and analysis program-national sampling design and estimation procedures. Gen. Tech. Rep. SRS-80. Asheville, NC: US Department of Agriculture, Forest Service, Southern Research Station. 85 p. 80.

Brown, T. N. and D. Kulasiri. 1996. Validating models of complex, stochastic, biological systems. Ecological Modelling 86: 129–134.

Cariboni, J., D. Gatelli, R. Liska and A. Saltelli. 2007. The role of sensitivity analysis in ecological modelling. Ecological Modelling 203: 167–182.

Congalton, R. G. 1991. A review of assessing the accuracy of classifications of remotely sensed data. Remote Sensing of the Environment 37: 35–46.

Congalton, R. G. and K. Green. 1999. Assessing the accuracy of remotely sensed data: Principles and Practices. Lewis Publishers, CRC Press, Seattle, WA.

Cox, D. R. and E. J. Snell. 1968. A general definition of residuals. Journal of the Royal Statistical Society. Series B (Methodological): 248–275.

Crookston, N. L. and G. E. Dixon. 2005. The forest vegetation simulator: A review of its structure, content, and applications. Computers and Electronics in Agriculture 49: 60–80.

Fall, J. G., B. Dorner and K. Lertzman. 1997. A model for quantifying uncertainty in fire history studies. p. 72. *In*: 78th Annual Meeting of the Pacific Division, American Association for the Advancement of Science (AAAS), Oregon State University, Corvallis.

Fawcett, T. 2006. An introduction to ROC analysis. Pattern Recognition Letters 27: 861–874.

Finney, M. A. 1998. FARSITE: Fire Area Simulator—model development and evaluation. Research Paper RMRS-RP-4, United States Department of Agriculture, Forest Service Rocky Mountain Research Station, Ft. Collins, CO USA.

Gardner, R. H. and D. L. Urban. 2003. Model validation and testing: Past lessons, present concerns, future prospects. *In*: C. D. Canham, J. C. Cole and W. K. Lauenroth (eds.). Models in Ecosystem Science. Princeton Univ. Press, Princeton, NJ USA.

Hamby, D. M. 1994. A review of techniques for parameter sensitivity analysis of environmental models. Environmental Monitoring and Assessment 32: 135–154.

Helmreich, J. E. 2016. Regression modeling strategies with applications to linear models, logistic and ordinal regression and survival analysis. Journal of Statistical Software 70.

Holsinger, L., R. E. Keane, D. J. Isaak, L. Eby and M. K. Young. 2014. Relative effects of climate change and wildfires on stream temperatures: A simulation modeling approach in a Rocky Mountain watershed. Climatic Change 124: 191–206.

Jakeman, A. J., R. A. Letcher and J. P. Norton. 2006. Ten iterative steps in development and evaluation of environmental models. Environmental Modelling & Software 21: 602–614.

Keane, R. E. and M. A. Finney. 2003. The simulation of landscape fire, climate, and ecosystem dynamics. pp. 32–68. *In*: T. T. Veblen, W. L. Baker, G. Montenegro and T. W. Swetnam (eds.). Fire and Global Change in Temperate Ecosystems of the Western Americas. *Springer-Verlag*, New York, New York, USA.

Keane, R. E., D. McKenzie, D. A. Falk, E. A. H. Smithwick, C. Miller and L.-K. B. Kellogg. 2015. Representing climate, disturbance, and vegetation interactions in landscape models. Ecological Modelling 309-310: 33–47.

Keane, R. E., P. Morgan and S. W. Running. 1996. FIRE-BGC—a mechanistic ecological process model for simulating fire succession on coniferous forest landscapes of the northern Rocky Mountains. Research Paper INT-RP-484, United States Department of Agriculture, Forest Service Intermountain Forest and Range Experiment Station, Ogden, UT USA.

Keane, R. E., R. Parsons and P. Hessburg. 2002. Estimating historical range and variation of landscape patch dynamics: Limitations of the simulation approach. Ecological Modelling 151: 29–49.

Kleijnen, J. P. C., G. v. Ham and J. Rotmans. 1992. Techniques for sensitivity analysis of simulation models: A case study of the CO_2 greenhouse effect. Simulation 58: 410–417.

154

Krieger, T. J., C. Durston and D. Albright. 1978. Statistical determination of effective variables in sensitivity analysis. Transactions of the American Nuclear Society 28.

Liu, S. G., B. Bond-Lamberty, J. A. Hicke, R. Vargas, S. Q. Zhao, J. Chen, S. L. Edburg, Y. M. Hu, J. X. Liu, A. D. McGuire, J. F. Xiao, R. Keane, W. P. Yuan, J. W. Tang, Y. Q. Luo, C. Potter and J. Oeding. 2011. Simulating the impacts of disturbances on forest carbon cycling in North America: Processes, data, models, and challenges. Journal of Geophysical Research-Biogeosciences 116.

Lorscheid, I. and M. Meyer. 2016. Divide and conquer: Configuring submodels for valid and efficient analyses of complex simulation models. Ecological Modelling 326: 152–161.

Lutes, D. C., N. C. Benson, M. Keifer, J. F. Caratti and S. A. Streetman. 2009a. FFI: a software tool for ecological monitoring*. International Journal of Wildland Fire 18: 310–314.

Lutes, D. C., R. E. Keane and J. F. Caratti. 2009b. A surface fuels classification for estimating fire effects. International Journal of Wildland Fire 18: 802–814.

Lutes, D. C., R. E. Keane, J. F. Caratti, C. H. Key, N. C. Benson, S. Sutherland and L. J. Gangi. 2006. FIREMON: Fire effects monitoring and inventory system. General Technical Report RMRS-GTR-164-CD, USDA Forest Service Rocky Mountain Research Station, Fort Collins, CO USA.

Oreskes, N. and K. Belitz. 2001. Philosophical issues in model assessment. Model Validation: Perspectives in Hydrological Science 23.

Oreskes, N., K. Shrader-Frechette and K. Belitz. 1994. Verification, validation, and confirmation of numerical models in the earth sciences. Science 263: 641–646.

Orsi, F., D. Geneletti and A. C. Newton. 2011. Towards a common set of criteria and indicators to identify forest restoration priorities: An expert panel-based approach. Ecological Indicators 11: 337–347.

Parsons, R. A., E. K. Heyerdahl, R. E. Keane, B. Dorner and J. Fall. 2007. Assessing accuracy of point fire intervals across landscapes with simulation modeling. Canadian Journal of Forest Research 37: 1605–1614.

Piñeiro, G., S. Perelman, J. P. Guerschman and J. M. Paruelo. 2008. How to evaluate models: Observed vs. predicted or predicted vs. observed? Ecological Modelling 216: 316–322.

Pontius, R. G., O. Thontteh and H. Chen. 2008. Components of information for multiple resolution comparison between maps that share a real variable. Environmental and Ecological Statistics 15: 111–142.

Rollins, M. G., T. W. Swetnam and P. Morgan. 2001. Evaluating a century of fire patterns in two Rocky Mountain wilderness areas using digital fire atlases. Canadian Journal of Forest Research 31: 2107–2133.

Rykiel, E. J., Jr. 1996. Testing ecological models: The meaning of validation. Ecological Modeling 90: 229–244.

Saltelli, A., M. Ratto, T. Andres, F. Campolongo, J. Cariboni, D. Gatelli, M. Saisana and S. Tarantola. 2008. Global sensitivity analysis: The primer. John Wiley & Sons.

Schaefer, K., C. R. Schwalm, C. Williams, M. A. Arain, A. Barr, J. M. Chen, K. J. Davis, D. Dimitrov, T. W. Hilton, D. Y. Hollinger, E. Humphreys, B. Poulter, B. M. Raczka, A. D. Richardson, A. Sahoo, P. Thornton, R. Vargas, H. Verbeeck, R. Anderson, I. Baker, T. A. Black, P. Bolstad, J. Chen, P. S. Curtis, A. R. Desai, M. Dietze, D. Dragoni, C. Gough, R. F. Grant, L. Gu, A. Jain, C. Kucharik, B. Law, S. Liu, E. Lokipitiya, H. A. Margolis, R. Matamala, J. H. McCaughey, R. Monson, J. W. Munger, W. Oechel, C. Peng, D. T. Price, D. Ricciuto, W. J. Riley, N. Roulet, H. Tian, C. Tonitto, M. Torn, E. Weng and X. Zhou. 2012. A model-data comparison of gross primary productivity: Results from the North American Carbon Program site synthesis. Journal of Geophysical Research: Biogeosciences 117.

Smaldino, P. E. 2013. Measures of individual uncertainty for ecological models: Variance and entropy. Ecological Modelling 254: 50–53.

Sokolova, M. and G. Lapalme. 2009. A systematic analysis of performance measures for classification tasks. Information Processing & Management 45: 427–437.

Stage, A. R., C. R. Hatch, D. L. Rice, D. W. Renner, J. J. Coble and R. Korol. 1995. Calibrating a forest succession model with a single-tree growth model—An exercise in meta-modeling. pp. 194–209. *In*: Recent Advances in Forest Mensuration and Growth Research. Danish Forest and Landscape Research Institute. Tampere Finland.

Vihinen, M. 2012. How to evaluate performance of prediction methods? Measures and their interpretation in variation effect analysis. BMC Genomics 13: S2–S2.

Xu, C., Y. Hu, Y. Chang, Y. Jiang, X. Li, R. Bu and H. He. 2004. [Sensitivity analysis in ecological modeling]. Ying yong sheng tai xue bao = The Journal of Applied Ecology 15: 1056–1062.

Zar, J. H. 1996. Biostatistical Analysis. Third edition. Prentice Hall, Upper Saddle River, NJ.

8

Execution

Implementing the Model Project

.

"Everyone has a plan: until they get punched in the face."

Mike Tyson

Execution—the carrying out or putting into effect of a plan, order, or course of action (OED).

ABSTRACT

This chapter covers how the model is used to actually implement a simulation design in a modeling project. In the purest sense, "execution" is the running of the model, but here, the term is used in a broader sense—running the model for all replicates in the simulation design. There are five simple steps involved in project execution: (1) understand the model to know if there is a problem, (2) find computing resources, (3) design output storage structures, (4) develop computer scripts to efficiently run the model for all scenarios, and (5) start the simulations and scripts. Major execution concerns are addressed at the end of the chapter.

157

Introduction

In this chapter we outline how the model is used to actually implement a modeling project. In the purest sense, "execution" is the running of a model, but in this book, the term is used in a broader sense—the execution phase involves running the model for all replicates in the simulation design. As an example, the execution phase may involve 120 executions of the model capturing four scenarios each with three levels and 10 replicates. This phase is by far the easiest and sometimes the quickest task in the entire project, and as a result, this chapter is somewhat short. However, there are a few aspects of executing a model that deserve mention.

Execution Steps

Once the calibration is done (Chapter 6), and the user or modeling team feels that the initialization (Chapter 4) and parameterization (Chapter 5) are sufficient, then the model is ready for the execution of simulations in the modeling project design (Chapter 3). It is advised that the validation phase of the modeling project is complete (Chapter 7), but it need not be finished to conduct the execution of the model. The four steps involved in execution are simple but important (Figure 8.1).

Understand model

At this point in the modeling project, the user should have comprehensive knowledge of the details of the selected model including input/output requirements, important algorithms, and of course, the procedure needed to execute the model. It is critical that the user has read all publications about the model and understands many of the nuances of the selected model. If this is difficult, as it is when someone is new to the modeling field, then it is important to cultivate relationships with not only the modeler and developer, but also with other people that have used the model and other modelers (see next section). This is also important because the user needs to understand the modeled output for the project to properly implement the analysis phase (see next section).

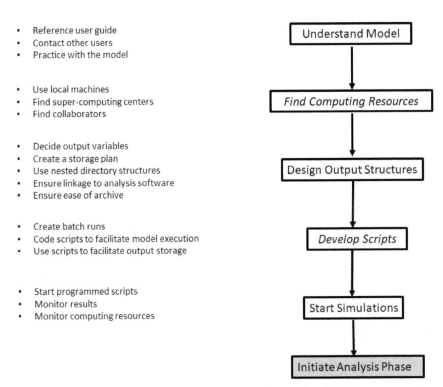

- Reference user guide
- Contact other users
- Practice with the model

- Use local machines
- Find super-computing centers
- Find collaborators

- Decide output variables
- Create a storage plan
- Use nested directory structures
- Ensure linkage to analysis software
- Ensure ease of archive

- Create batch runs
- Code scripts to facilitate model execution
- Use scripts to facilitate output storage

- Start programmed scripts
- Monitor results
- Monitor computing resources

Understand Model

Find Computing Resources

Design Output Structures

Develop Scripts

Start Simulations

Initiate Analysis Phase

Figure 8.1. The steps used in the execution phase of a modeling project.

Users should practice running the model before starting the execution phase. Most modelers provide example input structures for easily executing their model to see how it works. Prior to or during the previous three phases (initialization, parameterization, and calibration), novice users can take these example input files and execute the model to become familiar with the model and gain the confidence to know when the model is working correctly. Then they should modify various parameters in the example input files to learn model behavior. By practicing, the user can make a more informed estimate of how long the model will take to simulation their project, and understand when the model is performing correctly.

159

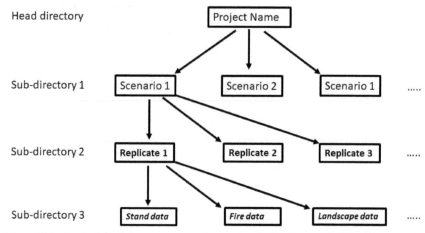

Figure 8.2. A possible directory structure for storing output from a modeling project. The name of the project is the head director with subdirectories organized first by scenarios, then by replicates. then by the type of data file. All of these files can be put in one directory but their filenames should reflect the simulation design.

Find computing resources

If the selected model is somewhat simple and executes in seconds to minutes on your laptop computer, then the reader can bypass this step and not worry about cobbling together computer resources to complete the project. However, most modeling projects that I've worked on were limited by available computing resources such as number of computers, configuration of processors, available memory, and CPU speed. Often, complex models take hours to days to run; may demand extensive memory and computing processing; and could produce tremendous amounts of outputs. This means that the user must find a computing "platform" that can handle all computing requirements of the project.

The user should first determine how long it takes for the model to run a given scenario replicate on a "local" computer (i.e., the user's computer or local server). Some ecological models may take hours to finish one run; the FireBGCv2 model, for example, took 18 hours to simulate 400 years on a 100,000 ha landscape. Once the execution time on the local computer is known, the user can then do the math to compute how long it will take to do all scenarios and replicates. If the estimated time is too long then compromises must be made to complete the project within acceptable timelines, and the user must develop a computing plan. This involves finding other local computers to handle the load or finding other computing resources outside the project domain. Supercomputing resources, for example, are available on the web and government agencies and university organizations have supercomputing facilities. Purchasing computers specifically for the project may also be an option.

It would be beyond the scope of this book to detail all possible options and procedures to find and implement the computing resources needed for a specific modeling project. However, here are some tips and suggestions that may make project execution go easier:

1. **Ensure extensive disk storage.** The output generated by some models can be overwhelming. It is important to have enough disk storage to hold all output for all scenarios and replicates. Buying additional USB disk storage is a must for some projects.

2. **Revisit project design.** If the user finds out that there are insufficient computing resources to effectively execute the project and it would take substantial time and money to find these resources, perhaps the next step should be to see if the project design can be modified so that the available resources can handle the full implementation of the project. Are all the replicates needed? Can one scenario be eliminated? Can fewer years be simulated? However, users should thoroughly document all changes to the project implementation so results can be interpreted in the correct context.

3. **Identify computing resources beforehand.** A concerted effort should be used to find all available computers during the project design phase rather than waiting until the model is ready to be actually executed. This may be difficult because details of the simulations are unknown, but at least a preliminary inventory will identify if outside resources are indicated.

4. **Interview other users.** It often helps if the user contacts other users of the same model to determine the appropriate amount of computing resources needed for project implementation and to gain insight into possible solutions.

5. **Foster relationships with IT folks.** A healthy relationship with the resident computing staff at an organization is critical if the modeling project is to go off without a hitch. They will often have creative ideas on how to solve many of the unexpected problems involved in executing the model for the best results. Cloud computing, for example, might be an alternative to simulating the model on local computers.

Design output structures

The analysis phase (Chapter 9) of a modeling project will go much easier if the user has the forethought to design a comprehensive storage plan for the simulation results. This often involves designing a directory structure that stores modeled output in an understandable and accessible way so that analysis software can easily access these files for speedy investigation. The data should be easily accessible in a completely understandable format for sharing across the modeling team. For example, the top directory could be named for the project, the sub-directories named for the scenarios, and their sub-directories named for the replicates (Figure 8.2).

The most important step in this design structure is to revisit the selected variables to output from the model (Chapter 3). Now that the project is ready to be executed and all other preliminary phases have been completed (initialization, parameterization, and calibration), there may be additional variables that have been identified as important from analyses in previous phases, especially the calibration phase. Many projects had to redo their entire execution phase after it was revealed that an important variable was forgotten. Revisit the list of variables and have the modeling team review the list just to make sure all is ready.

Develop scripts

Most users would like to start the model and walk away while it simulates all of the identified scenario replicates (Chapter 3). It might take a model four hours, for example, to simulate one replicate in the project's simulation design, so most people don't want to sit around and wait for the model to be finished before starting the simulation of another run. Therefore, it is critical that the user learn how to write scripts to run all scenarios in one "batch" run. Most operating systems provide the ability to execute multiple runs of a single or multiple programs using a script programming language. Sometimes it is quite simple to learn the steps to program the execution of multiple runs of the model under the simulation design of the project.

Unfortunately, this section cannot provide the information needed to program scripts in the multiple operating systems currently available on most computer systems, such as MS-DOS, UNIX, and LINIX; users should reference the manuals for those systems. However, there are a few suggestions that might make the process easier. First, be sure the directory structure is integrated into the scripts so that the input/output data are easily available. Next, extensively test the script and then test it again. Try to avoid any potential problems by evaluating results of the script. And last, consult a software expert in the resident operating system to ensure your scripts are appropriate and effectively written.

Start Simulations

Results of all the hard work performed during the first three phases are now realized as the model is executed for all replicates. This step always seems anticlimactic because it takes the least amount of work, but many users have breathed a sigh of relief once the runs are started. To implement this step, the user simply executes the model or the programmed script. It is important that the first few runs of the model are monitored to determine how long the executions are taking and to ensure there are enough hardware resources (e.g., memory, nodes, processor speed) to handle the model executions.

Execution Concerns

In my experience, no project has ever been completed with only one execution. Mistakes are always made in the design and implementation of the parameterization, initialization, and calibration phases that become evident when the execution of the entire project is finished and the user starts the analysis phase. As a result, the model must be re-executed for all scenarios and replicates. This is somewhat problematic as simulation times on some of the projects I worked on were 1–2 months. There are several ways to mitigate the impact of this on project success. First, it is critical that the user evaluate results of the initialization (Chapter 4), parameterization (Chapter 5), and calibration (Chapter 6) in the context of the project's objective to ensure everything is working properly. For example, the user should augment the project's output variables with other exploratory variables to ensure that all output variables can answer the objective. Next, in any idle moments prior to project execution, the user can run the model *ad hoc* to prototype and validate the output structures (previous section) and also to evaluate the output for acceptability. Also, the user can thoroughly evaluate the quality of model output. Many times I've noticed one mistake in the model output and fixed that mistake only to notice yet another mistake after the entire project's simulation was completed. And last, the user can extend the project's timelines to include at least two full model executions of the project design. Regardless of how many safeguards and pro-active controls that were put in place to avoid running the model for the entire project again, some modeling projects will need to re-execute the model after mistakes are noticed. The user should expect this and hope that the project's simulations need only two attempts.

Many ecosystem models do not have the sophisticated error handling and documentation that are common in commercial software packages such as spreadsheets, word processors, and statistics packages. As a result, it is often difficult for users to find exactly what went wrong during a simulation if the model stopped unexpectedly or it produced odd or erroneous output. Therefore, it is important that the user develop safeguards to make sure model bugs, system updates, hardware failures, and memory overruns are avoided. First, the user could attempt to "break" the model by running the model with odd parameters. For example, the user could try running the model for 10,000 years to determine if it overwhelms computer memory. Or the user could try inputting a parameter value that is illogical or ecological invalid to see how the model accepts and uses this abnormality. Next, the user should try to find someone who may be familiar with the model so that they can help debug program errors (see previous sections).

9

Analysis

Evaluating Model Results

"If you torture the data long enough, it will confess."

Ronald Coase, Economist

Analysis—a detailed examination of anything complex in order to understand its nature or to determine its essential features (Merriam Webster 2018).

ABSTRACT

Ecological models often output terabytes of data that must be synthesized and summarized into salient facts to answer the modeling objective. This involves advanced statistical analysis of the simulated data in a spatial and non-spatial context to determine those significant differences between scenarios that will answer the questions raised by the modeling objective. First, this chapter covers the four steps of analysis: (1) decide on analysis software, (2) link output to analysis package, (3) perform analysis, (4) create final set of figures, tables, and maps. Then, various analysis issues are covered. Both response and explanatory variables are presented and various analysis summaries are presented.

Introduction

Many people think that once the model has been executed for all of the project's scenario replications, their work is done. Unfortunately, there is still yet another important step. Complex ecological models often produce terabytes of output data that must be now be synthesized and summarized into salient figures and tables that help answer the modeling objective. This often involves advanced statistical analysis of the simulated data in a spatial and non-spatial context to determine those significant differences between scenarios that will answer the questions raised by the modeling objective (Zuur et al. 2007). Moreover, it is important that the modeling team understand "why" a particular result occurred, therefore, additional analyses if often needed to summarize the simulated data to understand the reasons for a particular result.

The most important goal of any analysis of simulated data in a modeling project is to answer the objective (Chapter 3). If there was a clear statement of objectives for the modeling project and the modeler implemented the steps outlined in Chapter 3 (namely prototyping tables and figures needed for successful completion of the objective), then the analyses of the model output data should be rather straightforward. However, I have never worked on a modeling project when the modeling objective didn't change at some point during the project; often preliminary evaluations of the generated output reveal that a change in the project objectives was needed. It is extremely difficult to anticipate every possible result that the model will generate *a priori* (i.e., beforehand). Often, there will be simulation results that were unexpected which precipitates the need for further analyses to determine why the results seem odd. In most cases, explanatory variables are evaluated against response variables that create numerous series of graphs, tables, and statistical summaries to provide the user with the synthesized information needed to understand simulation results.

As mentioned in Chapter 3, model results can be divided into two categories—response variable and explanatory variable output (GRIMM 2002). Response variables are used to address the modeling objectives while explanatory variables are used to figure out why the model produced the unique results for those response variables. As an example, say that

166

the modeling objective was to determine the fate of a tree species on a landscape under two climate scenarios. The response variables might be a set of abundance measures, such as basal area, tree density, and leaf area by species. However, these response variables may not provide insight as to why a tree species' abundance trended down or up—explanatory variables, such as climate (e.g., rainfall, evapotranspiration), disturbance (e.g., area burned by fire), and management (e.g., area thinned and prescribed burn) may provide insight into understanding the trends and explain how much each process drives a particular response. In a simple example, area by cover type is often reported from S&T models over simulation time, but to fully understand why a cover type increases or decreases, it is important to compare simulated cover type area with the area perturbed by fire and other disturbances.

It always seemed strange to me that high powered hypothesis-testing statistics were needed to analyze "simulated" data. These data were generated by an imperfect computer model that by definition would have a high amount of uncertainty and bias in the output that is impossible to capture in any statistical test. Moreover, it is relatively easy to generate more data if additional statistical power is needed by simply running the model a couple of more times (White et al. 2014). Statistics are wonderful tools for summarizing and comparing data, but they are not the final word when interpreting simulation output. Nothing takes the place of the human mind for understanding the subtle differences between modeled scenarios and incorporating known sources of uncertainty into the interpretation of the results. Simple box-and-whisker diagrams are sometimes more understandable, informative, and decisive than the most complex ANOVA analyses. Some users often stop their analyses when ANOVA tests for the factorial design are finished, but I feel that the ANOVA is only the start of the analysis and the really interesting part of a modeling project—interpreting modeled output—comes next.

Conversely, it is important that the primary user, in their zeal to understand simulated results, try to avoid going down the "rat holes" of tangential analyses. Users often get so enthralled with the minutia of investigating simulation results that they forget to concentrate on the project's objectives and get lost going down analysis pathways that are interesting but not important. To avoid this "scope creep", it is important that the user constantly pay attention to several things. First, the user must ask "does this analysis help answer the simulation objective?" Next, users should ask "do I have the time to conduct this ancillary analysis?" And last, they should ask "how can I integrate these analysis results into a report for the modeling project?" Too often, productive analysts become lost in interesting explorations because they didn't sufficiently understand the project's goals and products and they had a great fascination attempting to find unanticipated findings from the dense modeling output.

This chapter presents the basics of analyzing simulation data for the modeling project. Obviously, details of any analysis are unique to the model, the modeling objective, and the output generated, but the following are some general steps to guide model output analysis.

Analysis Steps

Decide on statistical software

Almost all simulation analyses tasks are accomplished using standard statistical analysis packages such as R (Team 2017), SAS (Inc. 1999), SPSS (SPSS 1999), and SYSTAT (Wilkinson 1988). In this book, all analyses are assumed to be accomplished using statistical software and that this software has the capability to efficiently synthesize the data, perform complex statistical analysis, and create high quality figures, tables, and graphs. Hopefully, the same software was used in previous modeling phases, especially validation and calibration, and therefore, this material would be relevant to those chapters as well.

The challenge, then, is to pick the most appropriate statistical software for analyzing the copious output from a modeling project. While the list of criteria for software selection may be quite long for some projects, I've found that there are three major factors that are most important in many projects. First, the user should identify what statistical packages are available from their organization that are within their budget—in other words, which packages are already resident on their computer or available for free. Most modeling projects can't afford to purchase a brand new high end statistical package, so other low cost alternatives are often sought. Next, the selected software package must be able to ingest the abundant simulated data and perform the desired analyses. I've participated in several modeling projects when the great amount of simulated data overwhelmed the computing ability of the program (i.e., the analysis took too much time). And last, there must be someone available within the user's organization or nearby that can answer questions on the software's useage and analysis results.

Link data to analysis package

The next step is to *link* the output files to the analysis package (Figure 9.1). This might involve "massaging" the data to fit the input requirements of the analysis software, or programming the software to accept the data in its native output format. The formats of the output data are readily identified

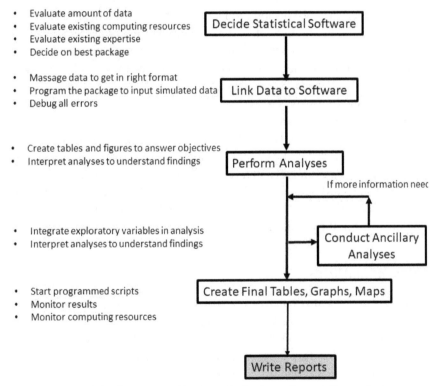

Figure 9.1. The steps used in the analysis phase of a modeling project.

from the models user's guides or programming code; the simulated output, for example, may be stored as comma-delineated fields in ANSI-ASCII text files. But some models save output in special formats, such as binary, netCDF, or geotiff, which may be inaccessible to some statistical packages. Additional steps must then be taken, with additional software packages such as database management packages (e.g., Oracle, Access), C++ programs, and free-ware programs.

To ensure that input data are error-free and ready for analysis, it is recommended that a set of summary tables, figures, and data queries be created using the statistical software to scrutinize the data for abnormalities. Summary tables and figures that display results over time and space can be used to ensure the data are being read into the package correctly and to determine if there are any irregularities, outliers, or simulation errors. Sometimes, during model execution, the computer may malfunction and spurious values may be written to an output file that fall outside the bounds of acceptable conditions—such as a negative value or an alpha character in a numeric field. Data queries using a text editor can be used to find and fix these problems or they can be excluded from the analysis using the statistical packages. However, it may also be necessary to explore the possibility that model input or parameterization are responsible for the outliers, or in more exceptional cases, the inherent model design is the problem, such as the model was not constructed for the particular characteristics of the simulation landscape or its ecological components (e.g., topographic, climate, vegetation). Such issues are usually discovered and resolved in earlier steps of the modeling process, but some complications may only come to light when the model is fully implemented in the simulation phase.

Perform analyses

In this step, the statistical software package is used to conduct analyses on the output needed to answer the objective. If the user has completed the tasks mentioned in Chapter 3, namely designed the figures, tables, and statistical tests, then this step is relatively easy. But as mentioned, analyses usually bloom into myriad investigative pathways and this can lead to new and often interesting areas. It is suggested that analysis efforts first focus on answering the project's main objectives, which may then set the context for additional analysis that may be necessary to understand the key results. The purpose of this chapter is NOT to prescribe a series of analysis tasks and statistical tests because those are dictated by the project's objective which is highly unique. The purpose of this chapter is to suggest some common analysis techniques and products that are common to many modeling projects, and these are presented by the output products.

There are basically three types of products that are generated from simulation data during the analysis phase—tables, figures, and maps. Tables contain the summaries of complex statistical analysis for reference in report writing and are usually selected when the actual numbers summarized in the table will be cited by others for other purposes. Figures are illustrative diagrams that summarize and synthesize simulated data for results interpretation and report writing. And maps are georeferenced displays of the spatial distribution of modeled entities. Suggested analyses are presented by each of these products.

Tables

There are a number of statistical tables that might be considered for the interpretation of simulation results. First and foremost, results of a parametric (ANOVA) or non-parametric significance test of differences of response variables across simulated scenarios should be considered (Table 9.1). The simulated data should be first examined to see if they are normally distributed so that parametric statistics can be used, and if not normal, then non-parametric tests may be appropriate. After many consultations with statisticians, it appears that ANOVA with repeated measures (e.g., across replicates and years) is the preferable approach to detect differences in response variables between scenarios for time-series simulation data that are normally distributed. For non-normal data, a generalized linear mixed model (GLMM) for repeated measures is a complementary and useful tool (Table 9.1) (Bolker et al. 2009). In many projects, the number of observations can be boosted by using years (or time intervals) across replicates (e.g., a 200 year simulation with results reported every 10 years with five replicates can generate 20 x 5=100 observations). However care must be taken in interpreting results because

Table 9.1. Results of generalized linear mixed models for the effects of planting, restoration, suppression climate and year for the amount of whitebark pine basal area and the proportion landscape area in whitebark pine dominated communities in the whitebark pine zone of the East Fork of the Bitterroot River (EFBR) for Future only, Historic only, and Future and Historic combined. Asterisk indicates significant interaction between effects. From Keane et al. (2017).

Effect	EFBR		
	F	**Degrees of freedom** Numerator; denominator	**P-value**
Whitebark pine basal area (m^2 ha^{-1})			
Future			
Restoration	145.80	2; 263	< .0001
Planting	0.98	2; 263	0.38
Suppression	2.05	2; 263	0.13
Year	760.92	9; 2421	< .0001
Historic			
Restoration	31.89	2; 259	< .0001
Planting	0.70	2; 259	0.50
Suppression	1.55	2; 259	0.21
Supp X Restore	3.69	4; 259	0.006
Year	894.39	9; 2421	< .0001
Future and Historic			
Restoration	130.97	2; 532	< .0001
Planting	0.03	2; 532	0.97
Suppression	2.29	2; 532	0.10
Climate	234.11	2; 532	< .0001
Climate X Year	356.83	9; 4842	< .0001
Year	1274.36	9; 4842	< .0001

with high power (i.e., many observations), a significant effect might be detected that is actually unimportant (as discussed in more detail later in this chapter). Multiple comparisons, such as using Tukey's honestly significant difference (HSD) test, can also be conducted to evaluate whether response variable means are significantly different among scenarios (Table 9.2). The final analysis and its interpretation should also be reviewed by the local statistician if available.

Table 9.2. Multiple comparisons of the main treatment effect's least square means (± standard error) on whitebark pine (PIAL) basal area and proportion landscape area of the EFBR landscape for planting and restoration scenario levels. Tukey's honestly significant difference (HSD) test was used to evaluate whether means were significantly different from each other with corrections for unbalanced replication using the Tukey-Kramer. Future and historical climates evaluated separately and combined.

Variables	EFBR landscape			
	Restoration			
	None	Low	High	P-value†
Basal area (Future)	2.32 ± 1.03	4.77 ± 1.03	4.77 ± 1.03	a, < 0.0001
Basal area (Historic)	1.34 ± 1.04	2.08 ± 1.04	1.72 ± 1.04	b, < 0.0001
Basal area (Future/Historic)	1.77 ± 1.03	3.13 ± 1.03	2.83 ± 1.03	c, < 0.0001
Prop. PIAL (Future)	0.1217 ± 0.006	0.4558 ± 0.009	0.5025 ± 0.009	b, < 0.0001
Prop. PIAL (Historic)	0.0642 ± 0.003	0.1271 ± 0.004	0.1605 ± 0.004	d, < 0.0001
Prop. PIAL (Future/Historic)	0.0901 ± 0.004	0.2772 ± 0.006	0.3210 ± 0.007	d, < 0.0001
	Planting			
	None	Low	High	P-value†
Basal area (Future)	3.71 ± 1.03	3.86 ± 1.03	3.60 ± 1.03	n.s.
Basal area (Historic)	1.72 ± 1.04	1.62 ± 1.04	1.72 ± 1.04	n.s.
Basal area (Future/Historic)	2.51 ± 1.03	2.51 ± 1.03	2.48 ± 1.03	n.s.
Prop. PIAL (Future)	0.3177 ± 0.009	0.3380 ± 0.009	0.3302 ± 0.009	n.s.
Prop. PIAL (Historic)	0.1097 ± 0.003	0.1098 ± 0.003	0.1116 ± 0.004	n.s.
Prop. PIAL (Future/Historic)	0.2023 ± 0.006	0.2112 ± 0.006	0.2090 ± 0.006	n.s.

Often, users find it difficult to decide if the output should be summarized in a table or figure. A general rule of thumb is that if the values of the summary statistics will be used by others in the project or organization, or if statistics are important to demonstrate the findings of a project, then a table is warranted. However, if a general trend, relationship, or association is needed to illustrate aspects of the modeling output, then figures are probably better than tables. As an example, if the historical range and variation (HRV) of wildfire burned area is needed to illustrate changes in vegetation types, then a box-and-whisker plot might be all that is needed, however, if the HRV of burned area is needed by the planning team to implement a particular scenario, then the results should be summarized in a table.

Figures

Box-and-whisker plots of each response variable across scenarios provide perhaps the best visual assessment of differences across scenarios (Figure 9.2). The vertical length of the box and the whisker identify the range and percentiles in the data (often the 10th and 90th for the whisker and 25th and 75th for the box) and the line in the box depicts a central tendency statistic (mean, mode, or median). Differences between scenarios can be easily visually assessed using this technique, but there are probably numerous other figure designs that could also be used. The primary challenge in designing a figure that accurately portrays simulation results is to show major differences in scenarios and their various levels across all factors (i.e., depicting multidimensional relationships).

The main problem with box and whisker plots is that they portray only ONE response variable, and often the differences across scenarios and their levels will differ by which response variable is used. For example, the box and whisker plots for basal area (Figure 9.2) may be completely different when timber volume or fuel loading is used (Keane et al. 2018). Many users want to combine all response variables together in a single analysis to determine differences between scenarios. Various statistical procedures are available to perform this task, specifically Principal Components (PCA) (Keane et al. 2018) and non-metric multidimensional scaling (NMDS) (Kenkel and Orlóci 1986). In the PCA statistical procedure, values

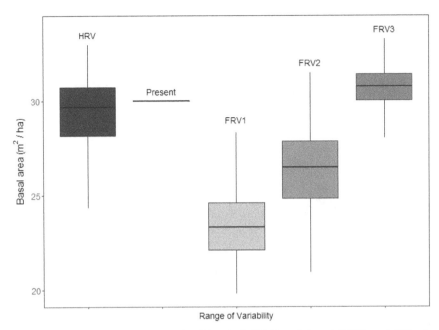

Figure 9.2. An illustration comparing historical (HRV) and future (FRV) variability in basal area (m^2 ha^{-1}) variability compared with current conditions on the EFBR landscape (Present: the initial conditions at the start of the simulation). There appears to be a zone of overlap between HRV and FRV3, which may provide a possible reference for management. FRV1, FRV2, and FRV3 are future simulations with RCP4.5 climates with zero, 50% and 98% of fire ignitions suppressed respectively. The box in this figure are the 25th and 75th interquartiles and the whiskers represent the range of the data. From Keane et al. 2018.

of all response variables can be used to compute at least two principal components that hopefully explain over 50% of the variance. Then the cloud of points that are the component values for each replicate or year or both can be displayed and analyzed (Figure 9.3). Moreover, the values of these principal components can then be used to create box and whisker plots across the scenarios and levels similar to the ones in Figure 9.2.

After response variables have been analyzed, it's time to start analyses of the explanatory variables to determine why there are differences across scenarios or not. And, perhaps the best first step is to create time series graphs of all response variables (i.e., values of response variables over time) (Figure 9.4) and repeat this for the explanatory variables (Figure 9.5). Rapid changes in response variables can be compared with

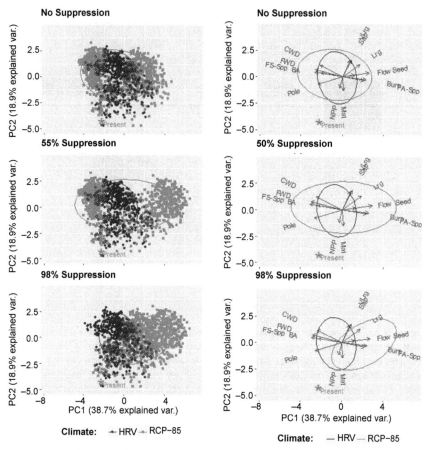

Figure 9.3. Results of PCA analysis of FireBGCv2 simulations for the EFBR landscape for the historical scenario (HRV; black dots, reference) and for the three future scenarios (FRV1, FRV2, FRV3; gray dots; Table 2) which are three fire suppression scenarios (no suppression, 50% ignitions suppressed, 98% ignitions suppressed) under an RCP8.5 climate (A,C,E). Also shown are simplifications of the scatterplots with circles that contain 68% of the variation in the spread of the points for the three FRV scenarios (B,D,F). The asterisk at the lower left of graphs A, C, E represents the condition of the landscape today (Present). FRV1 is shown in A and B, FRV2 is C and D, and FRV3 is E and F. Variable names in B, D, and F are defined in Table 3 and indicate the importance of the variables in the PC1 and PC2 scores.

the explanatory time series to see if they match in time. Moreover, the magnitudes and trends of response variables can be compared with the explanatory variables to detect any relationships. Examples provided in

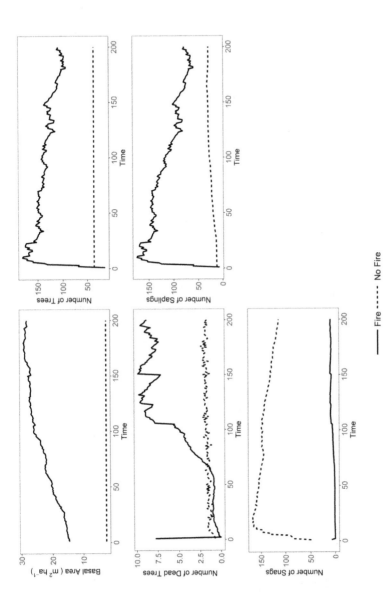

Figure 9.4. A set of graphs depicting the time series of response variables describing tree dynamics that include basal area, number of trees, number of recently dead trees, number of saplings, and number of snags (trees dead for more than one year) for the East Fork of the Bitterroot landscape (EFBR) from the FireBGCv2 model (Keane et al. 2011) for with fire (black line) and without fire (dotted line).

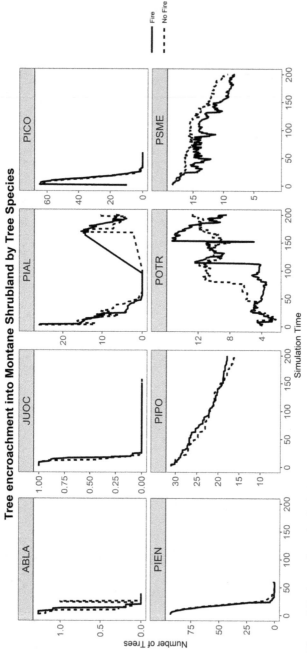

Figure 9.5. A set of graphs depicting the time series of explanatory variables that explain species changes in the tree dynamics response variables in Figure 9.4. Presented here are the number of trees by each tree species that occurred on a simulated plot for a montane shrubland in the East Fork of the Bitterroot landscape (EFBR) from the FireBGCv2 model for two scenarios (fire-black line, no fire-dotted line) (Keane et al. 2011). The abbreviations are as follows: ABLA-*Abies lasiocarpa* or subalpine fir, JUOC-*Juniperus occidentalis* or juniper, PIAL-*Pinus albicaulis* or whitebark pine, PICO-*Pinus contorta* or lodgepole pine, PIEN-*Picea engelmannii* or spruce, PIPO-*Pinus ponderosa* or ponderosa pine, POTR-*Populus tremuloides* or quaking aspen, PSME-*Psuedotsuga menziesii* or Douglas-fir.

Figures 9.4 and 9.5 show that the majority of trees in a montane meadow on the EFBR were juniper and quaking aspen. Of course, any detected correlation between response and explanatory variables using graphs and statistical techniques is not proof positive that there is a relationship. Only by digging deeper into the model structure can the actual relationship or feedback be revealed (i.e., a focused sensitivity analysis; Chapter 7). For example, the user may find that a response variable varies in tandem with a fundamental explanatory variable, such as aspen cover varies with area burned. Perhaps if box-and-whisker plots of the amount of burned area were compared with the basal area plots in Figure 9.2, the decline in basal area in the future may be interpreted as a result of the increases in wildland fire. However consider that a beetle outbreak may have preceded those large fires, and perhaps the outbreak was the actual cause of basal area loss. In short, it is important to consider a range of underlying mechanisms at a variety of temporal and spatial scales that might explain variation and trends in a response variable.

Another interesting explanatory analysis is to plot values of one variable against another across all years in a simulation and over all replicates (Figure 9.6). Here the user selects the more important response variables and plots them against explanatory variables to gain insight into model behavior and interpret model results. Different colors can be used for each scenario and scenario level.

Maps

Map products are only possible if the model was a spatial or landscape model and it contained options for outputting spatial data (Chapter 2). Some models can create digital maps of simulated output directly within the model structure, but most landscape models create files formatted for input into Geographical Information Systems (GIS) or other types of spatial software. Most maps created from simulation results simply present simulation results in a spatial display and rarely provide any spatial analyses (Figure 9.7). Many statistical packages, such as the R package (Team 2017), now have the ability to accept geo-referenced data to create colorful maps and perform important spatial analyses.

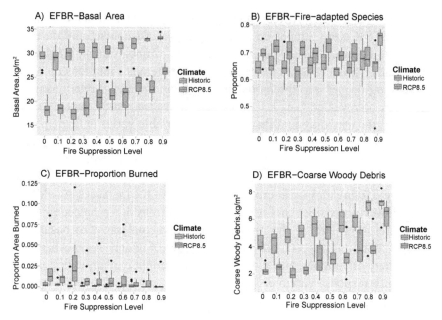

Figure 9.6. Graphs showing changes in four response variables over 10 fire suppression levels (proportion of fires suppressed) for historical (light boxes) and future (RCP8.5-dark boxes) climates for the East Fork of the Bitterroot River (EFBR, see Figure 2.9) generated by the FireBGCv2 landscape model. Response variables are: (A) average stand basal area (m^2 ha^{-1}), (B) proportion of landscape in fire adapted species, (C) average annual proportion of landscape burned each year, and (D) average coarse woody debris loading.

Most map products created from model output are usually for display purposes only, but some modeling projects conduct analysis on the map data to compute landscape metrics to describe landscape structure and composition. To perform these types of spatial data analysis, GIS and/or statistical software must be used to compute spatial statistics from output map data. Landscape metrics, such as contagion, average patch size, and juxtaposition, can be computed with landscape analysis software, such as FRAGSTATS (Baker and Cai 1992, McGarigal and Marks 1995, Dale 1999). Summaries of modeled output across geographic regions could be accomplished using GIS tools (Campbell et al. 1995, Akcakaya 1996). Spatial statistics, such as semivariogram analysis, Moran's I, and Geary's C (Cressie and Ver Hoef 1993, Fortin and Dale 2005), can also be computed using those statistical packages that include spatial options,

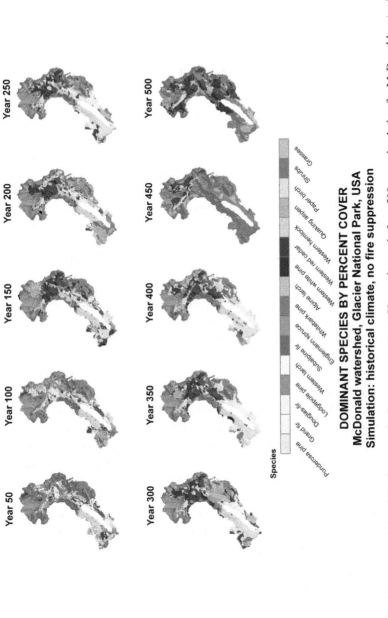

Figure 9.7. Map output of dominant species by percent cover at 50-year intervals for a 500-year simulation for McDonald watershed, Glacier National Park, USA (Keane et al. 2011).

such as R package (Team 2017). If these types of analyses are desired, the user should be aware that the output of digital maps from the model demands abundant disk storage.

Create Final Figures, Tables, and Maps

Once all initial and explanatory analyses have been completed to the mutual satisfaction of the modeling team, then it is time to create the necessary figures and tables to write the report that summarizes the results of the modeling project. This report may be a small synthesis of the simulation findings or a comprehensive paper to be submitted to a peer-reviewed journal. In any case, it is important that the final set of figures and tables captures the seminal findings of the modeling project in the context of the objectives, and also in the context of a greater understanding of ecosystem or landscape dynamics. A well-designed figure can successfully portray all results of an entire project and it will live forever if published.

High quality figures and tables should always be the goal of any analysis as they could be used or cited by any number of publications, especially if the results are novel and unanticipated. The challenge in this step is deciding which figures and tables need to be built. Too few figures may fail to accurately portray the depth and scope of the simulation results, while too many may be difficult to read, understand, and interpret. Here is where the creativity of the user and the modeling team comes into play as they will need to design figures and tables that accurately and parsimoniously synthesize simulation findings into an inspiring set of products. Without doubt, this is incredibly difficult for many of us as it is challenging to envision a design that will summarize the findings across the broad dimensionality of the abundant and diverse simulated output (i.e., scenarios, levels, and variables). Maps, figures and tables created in the preliminary analysis discussed in the previous section may be plentiful, simple, and crude, but final products must be designed so the reader easily understands the findings in the least number of products. The best way to find possible product designs is to peruse the literature on similar simulation projects and attempt to replicate their illustrative methods, or to review other reports generated by the host organization to find suitable designs.

Analysis Concerns

Statisticians often warn that there may be high spatial and temporal autocorrelation in simulated data that could influence many parametric statistical tests, such as analysis of variance (Strachan and Harvey 1996, Moreno-Fernández et al. 2016). Outputting spatial data across a landscape at individual time steps would obviously have unavoidable spatial and temporal autocorrelation in the data (Keane et al. 2006). But then again, all ecological data, empirical or simulated, have a certain amount of spatiotemporal correlation. Early modeling projects that I've worked on attempted to minimize spatiotemporal autocorrelation by outputting at long time intervals (> 50 years) and only reporting distant stands (> 5 km) (Holsinger et al. 2006, Pratt et al. 2006). These thresholds were identified by performing preliminary model runs for the selected landscape and outputting every year and then analyzing the time and spatial intervals that reflect the least temporal and spatial autocorrelation effects. However, some modelers feel that autocorrelation is part of the simulated system and it should be accepted as a component of natural variability so that important impacts and effects can be detected. For example, it may be that a 50 year reporting interval as used by Keane et al. (2006) was too long to detect significant impacts of frequent fire regimes in dry forest ecosystems or the prescribed burning programs implemented by management. Ultimately, the mitigation of the influence of spatiotemporal autocorrelation on modeled output is a decision of the modeling team in the context of the objective. Reporting every ten years, for example, may be needed to decide the success or failure of a simulated management alternative, but the analysis should recognize that there will be some temporal autocorrelation and decide if this autocorrelation matters to the overall result.

Simulation "oversampling" is often recognized as another challenge in modeling studies (White et al. 2014). In many projects, users employ parametric statistical tests of significance, such as ANOVA, to detect differences in simulated alternatives. However, the model that they used can be run for hundreds of years and scores of replicates to generate a profusion of observations to use in statistical testing. White et al. (2014) mention that this huge amount of observations may be inappropriate for two reasons: (1) p-values are calculated using great replication (i.e., statistical power), may produce significant p-values regardless of the effect size, and (2) the null hypothesis of no difference between treatments may be known *a priori* to be false thereby invalidating the premise of the test. They further state that "use of p-values is troublesome (rather than simply irrelevant) because small p-values lend a false sense of importance to observed differences." They argue that these analyses be abandoned and that the magnitude of differences between simulations be emphasized. In reality, there is no harm in performing the ANOVA to detect differences between alternatives, but the user should be aware of the effect of a high number of replicates on the significance test results, and perhaps create a design that keeps total observations less than 100–200. This is why it is always best to augment all ANOVA analyses with simple figures that show the magnitude of differences in response variables across scenarios.

Users should also be aware of using years as replicates (i.e., observations) in statistical significance tests such as ANOVA. The problem is that there are often trends in the simulated data that go undetected when response variables are summarized across all years. In one study, Keane et al. (2017) found few significant differences across modeled restoration scenarios, but there were significant declines over time in several climate change scenarios that were important to management actions that went undetected because the variability of the response variable was high because of the trends not the natural variablity of the processes. If years are used to boost the number of observations, be sure to also run significance test to evaluate the influence of years as an ANOVA factor or interaction.

Another analysis challenge for many users is deciding on the temporal span of analysis. A short temporal span may not detect long term differences across scenarios, and a span that is too long may be difficult to simulate with available computing resources. An example from Keane et al. (2017) depicts this conundrum. The statistical analysis for this project covered only the first 100 years of simulation (Figure 9.8) and the subtle differences across scenarios were a result of the rate of decline of whitebark pine. However, when a time span of 500 years was used, the impacts of climate and various management activities become manifest on the landscape and the differences between scenarios became evident. The small time slice was used because the climate projections only went out 100 years and this was the time span requested by managers, but ecosystem dynamics in this

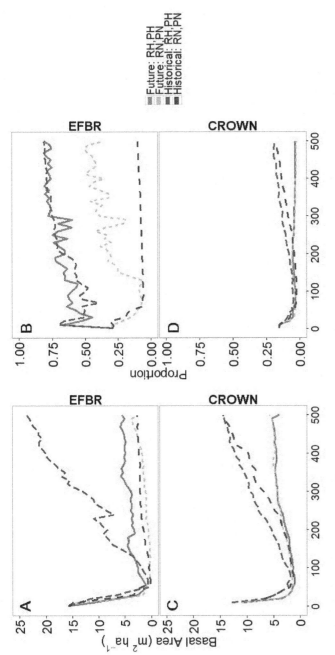

Figure 9.8. Results from the FireBGCv2 independent simulation runs from Keane et al. (2017). Increases in whitebark pine basal area in the future would occur on the higher elevation sites of the whitebark pine zone in the CROWN and EFBR landscapes under four scenarios (Future RH,PH-future climate and high restoration and planting activities-dotted line; Future RN,PN-future climate with no restoration and planting-solid line; Historical RH PH-historical climate with high restoration and planting; Historical RN PN-historical climate with no restoration and planting-dark dashed line). And whitebark pine on the CROWN landscape decline rapidly because there is more whitebark pine habitat at the lower elevations.

ecosystem are slow to develop and important changes weren't detected with such a small reporting interval.

The analysis phase is by far the most interesting and satisfying phase in the modeling effort. Users finally get to see the results of their extensive efforts expended during the previous phases. However, because this phase if often at the end of the modeling project, there rarely seems to be enough time to do a comprehensive analysis and acceptable exploration into the output data. Modeling projects without well-defined timelines are often delayed while unimportant analysis pathways are explored at the expense of report writing. To avoid over-analysis and timeline compression effect, users should remember the following:

1. Involve a statistician in designing the analysis products.
2. Be sure to answer the modeling objective.
3. Prepare all analysis tables, graphs, and maps with the intention that they will be included in the final report. Avoid creating poor quality graphs because of the compressed timelines.
4. Conduct a thorough review of the analysis results using the modeling team and modeling experts.
5. Use the experience gained from the previous phases in interpreting results.

References

Akcakaya, H. R. 1996. Linking GIS with Models of Ecological Risk Assessment for Endangered Species.

Baker, W. L. and Y. Cai. 1992. The r.le programs for multiscale analysis of landscape structure using the GRASS geographical information system. Landscape Ecology 7: 291–302.

Bolker, B. M., M. E. Brooks, C. J. Clark, S. W. Geange, J. R. Poulsen, M. H. H. Stevens and J.-S. S. White. 2009. Generalized linear mixed models: A practical guide for ecology and evolution. Trends in Ecology & Evolution 24: 127–135.

Campbell, J., D. Weinstein and M. Finney. 1995. Forest fire behavior modeling integrating GIS and BEHAVE. Ecosystem Management Center Report. USDA Forest Service, Washington D.C. USA.

Cressie, N. and J. M. Ver Hoef. 1993. Spatial statistical analysis of environmental and ecological data. pp. 404–413. *In*: M. F. Goodchild, B. O. Parks and L. T. Steyaert (eds.). Environmental Modeling with GIS. Oxford University Press, New York, NY., USA.

Dale, M. R. 1999. Spatial pattern analysis in plant ecology. Cambridge University Press, New York, New York, USA.

Fortin, M. J. and M. R. Dale. 2005. Spatial analysis: A guide for ecologists. Cambridge University Press, Cambridge, United Kingdom.

GRIMM, V. 2002. Visual debugging: A way of analyzing, understanding and communicating bottom-up simulation models in ecology. Natural Resource Modeling 15: 23–38.

Holsinger, L., R. E. Keane, B. Steele, M. C. Reeves and S. Pratt. 2006. Using historical simulations of vegetation to assess departure of current vegetation conditions across large landscapes. pp. 315–367. *In*: M. G. Rollings and C. Frame (eds.). The LANDFIRE prototype project: Nationally Consistent and Locally Relevant Geospatial Data for Wildland Fire Management. USDA Forest Service Rocky Mountain Research Station

Inc., S. I. 1999. SAS/STAT User's Guide Version 8.0. SAS Institute Inc., Cary, NC USA.

Keane, R. E., L. Holsinger, M. F. Mahalovich and D. F. Tomback. 2017. Restoring whitebark pine ecosystems in the face of climate change. pp. 123. USDA Forest Service Rocky Mountain Research Station, Fort Collins, CO.

Keane, R. E., L. Holsinger and S. Pratt. 2006. Simulating historical landscape dynamics using the landscape fire succession model LANDSUM version 4.0. General Technical Report RMRS-GTR-171CD, USDA Forest Service Rocky Mountain Research Station, Fort Collins, CO USA.

Keane, R. E., R. A. Loehman and L. M. Holsinger. 2011. The FireBGCv2 landscape fire and succession model: a research simulation platform for exploring fire and vegetation dynamics. General Technical Report RMRS-GTR-255, U.S. Department of Agriculture, Forest Service, Rocky Mountain Research Station, Fort Collins, CO USA.

Keane, R. E., R. A. Loehman, L. M. Holsinger, D. A. Falk, P. Higuera, S. M. Hood and P. F. Hessburg. 2018. Use of landscape simulation modeling to quantify resilience for ecological applications. Ecosphere 9: e02414.

Kenkel, N. C. and L. Orlóci. 1986. Applying metric and nonmetric multidimensional scaling to ecological studies: Some new results. Ecology 67: 919–928.

Lutes, D. C., R. E. Keane and J. F. Caratti. 2009. A surface fuels classification for estimating fire effects. International Journal of Wildland Fire 18: 802–814.

McGarigal, K. and B. J. Marks. 1995. FRAGSTATS: spatial pattern analysis program for quantifying landscape structure. General Technical Report PNW-GTR-351, USDA Forest Service.

Moreno-Fernández, D., L. Hernández, M. Sánchez-González, I. Cañellas and F. Montes. 2016. Space–time modeling of changes in the abundance and distribution of tree species. Forest Ecology and Management 372: 206–216.

Pratt, S. D., L. Holsinger and R. E. Keane. 2006. Modeling historical reference conditions for vegetation and fire regimes using simulation modeling. General Technical Report RMRS-GTR-175, USDA Forest Service Rocky Mountain Research Station, Fort Collins, CO USA.

SPSS. 1999. SPSS 10 for Windows. SPSS, Inc, Chicago, IL USA.

Strachan, I. B. and L. E. Harvey. 1996. Quantifying the effects of temporal autocorrelation on climatological regression models using geostatistical techniques. Canadian Journal of Forest Research 26: 864–871.

Team, R. C. 2017. R: A Language and Environment for Statistical Computing.

White, J. W., A. Rassweiler, J. F. Samhouri, A. C. Stier and C. White. 2014. Ecologists should not use statistical significance tests to interpret simulation model results. Oikos 123: 385–388.

Wilkinson, L. 1988. SYSTAT: The System for Statistics. SYSTAT Inc., Evanston, IL.

Zuur, A., E. N. Ieno and G. M. Smith. 2007. Analyzing ecological data. Springer Science & Business Media.

10

Issues

Things to Think About When Using Models

"Computers are good at following instructions, but not at reading your mind."

Donald Knuth

Modeling issues—broad concerns in successfully completing a modeling project.

ABSTRACT

There are a number of issues that occur before, during, and after a modeling project that will certainly influence any decisions and interpretations of the modeled results. Knowledge of these issues may make project design easier and may facilitate a more successful outcome. These issues are presented by several topics: (1) people, (2) climate, (3) disturbance, (4) historical simulations, and (5) modeling science. And last, a summary of this book is presented.

Introduction

In my experience, users of ecological models seem to be divided into three classes of people—those that think models are the answer to everything; those that think models are a necessary evil but are loathe to use them; and everyone else in between. The first class of people must temper their zeal for using ecological models by remembering that, at best, a model is an oversimplification of reality, and as a result, simulation results have a great deal of uncertainty that must be addressed during output interpretation. It is the user's job to identify the possible sources of uncertainty in model structure, parameterization, initialization, and output generation using concepts from this book and also to account for these sources in the understanding, interpretation, and use of model results.

The second class of people, the "model deniers", must remember that simulation modeling, while an odious task, may be the only method available to provide the best possible assessment of future impacts of management actions. Models are much better than an educated guesses or expert opinions for many applications (Gustafson et al. 2010). The advent of climate change, exotic invasions, and novel human land uses (e.g., fire exclusion) has made it difficult to assess future impacts using experience or empirical data because there are no historical analogs to understand novel ecological responses (Millar et al. 2007). Certainly, empirical approaches are often more desirable than some modeling approaches (Keane et al. 2015b), but we are now in the Anthropocene and the value of past phenomenological studies to assess management actions may be inappropriate for projecting into our new uncertain future (Gustafson 2013). Land management and planning professionals must face a future where the integration of aggressive modeling efforts with resident expertise, local knowledge, and empirical analysis are needed to ensure that the best available scientific information are used to plan and implement management actions (Schmolke et al. 2010).

All three classes of people must understand that every model has things that it does well and things that it does poorly. In the end, it is how a model is used and how the results are interpreted that are the most important, not the design or complexity of the model. It is essential that the strengths and weakness of any model be understood and addressed before that model is applied to a given situation. And, more importantly, it is important that users understand the assumptions made by the modeler during model construction. All information generated from the model can be useful depending upon the context in which it is interpreted. I've often found that the greatest thing learned during a modeling project is that the user often gains an understanding of great complexity of landscape and ecosystem dynamics from learning how to prepare, use, and interpret results from ecological models.

There are a number of important issues that occur before, during, and after a modeling project that will influence certain decisions and interpretations. Knowledge of these issues before embarking on a modeling project is started may make project design easier and may facilitate a more successful outcome. Some of these issues are not overly critical to the completion of a modeling project, but they do provide answers to questions that most novice users ask during a modeling project. These issues are presented by specific topics involved in a modeling project.

Common Issues

People

The most significant issue that I've encountered in my career is something I'd like to call **user's remorse.** As discussed, people that use other peoples' models, especially for applications not envisioned by the original modeler, often find the results somewhat unsatisfying, and often deem the results invalid or not useful. They look at the simulation findings through the lens of their own experience and come to the conclusion that the simulated outcomes are somehow unacceptable. This is in stark contrast to the reason that they decided to use a model in the first place—because they didn't know what would happen when a particular action was implemented. While a healthy degree of skepticism is an important part of the modeling process (see Chapter 6), over-criticism of a model and its results may work against the objectives of the modeling project and compromise interpretation of the results.

In a similar vein, some people in the modeling projects in which I've participated have often taken the position that the model results are "wrong" and it is up to the modeler or model must prove to them that the results are indeed acceptable. This is an example of a person using a model to tell them what they already know as opposed to a person who uses the model to obtain new information. Modelers are constantly reassuring skeptical users that the model is behaving correctly and the results are indeed valid. It may be best for these skeptics to remember that *all* model results are wrong, and, as a result, it is up to them to determine which results are the most useful (Box and Tiao 1975). I often wonder how many times the results from modeling projects have been thrown away because they were perceived as wrong, but in fact, were actually valid. In one modeling project that I conducted, a simulated stand of trees established after fire became stagnated after 50 years and did not self-thin as expected. The forester reviewing these results insisted that model was not behaving correctly because of his intense belief in the self-thinning rule. It turned out that the stagnation resulted from water limitations simulated on the dry forest landscapes, a phenomenon that was actually observed on the landscape being simulated. The point of this discussion is that modeling results can never be completely validated or fully assessed for accuracy, especially those results that project into the future. Modeling is just one tool in the user's bag of tools to make an educated guess about the impacts of management actions. Again, it is the user who decides how to interpret model results beneath the cloud of uncertainty.

Often, model users want to associate additional management values to modeled entities or variables to expand the domain of modeled results (**association**). An example is when the states in an S&T model (e.g., cover types) are assigned additional attributes, such as fuel loadings or grazing potential, to expand the scope of the simulation results (e.g., habitat suitability indices for grizzly bear are assigned to each state in the S&T model). This practice assumes two things: (1) the assigned attribute is indeed meaningfully correlated with that state, and (2) the states in the model accurately discriminate the range of attribute properties. Many times this practice can be quite successful and subsequent products, such as future maps of the new associated variables, can be easily validated using existing ground data. But often, the attribute values assigned to S&T states are inadequately represented by the resolution of state categories (Menakis et al. 2000). If users wish to associate attributes to modeled variables, it is essential that a validation analysis be conducted to ensure the assigned attributes are indeed distinctively discriminated by the simulated variable in question. For example, Keane et al. (2013b) performed a validation of a fuel classification and found that the classification categories did not adequately discriminate fuel loadings, yet these categories were being used throughout fire management. Similar validations of associated characteristics to modeled state variables are needed to establish the limits of uncertainty.

A last people issue is that many users often don't want to include a buffer around the target landscape because it may involve too much work or require too much simulation time. This is acceptable if the landscape model is not spatially explicit (i.e., does not include spatial processes such as disturbance spread, seed dispersal, animal movement). However, if a spatially explicit model is used and if the project proceeds without a buffer, users should know that there are real consequences to these actions, namely that the areas around the edges of the target landscape will not be fully impacted by spatial processes outside of the landscape and therefore will not be modeled the same rigor as the area in the interior. These areas can be cleaved from the analysis (i.e., a post-simulation buffer), or they can be included but the implications are incorporated into the results interpretation. In general, it is always better to include a buffer even if it adds additional time to the project.

Climate

A dilemma often encountered by ecological modelers is that the historical reconstructions of today's climates from GCMs or other finer scale climate models may inaccurately match observed weather (Taylor et al. 2012, Rupp et al. 2013). Comparisons of simulated to observed daily maximum temperature revealed that temperature was under-predicted by over 3°C for a landscape simulation in the US northern Rockies (Keane et al. 2017) and historical predicted and observed weather for the EFBR landscape did not mesh with future predictions (Figure 10.1). Climate modelers sometimes suggest using an offset method to correct this uncertainty (i.e., calculate an offset factor from a summary of observed vs predicted weather), and this works well for many projects. But, climate modelers also suggest

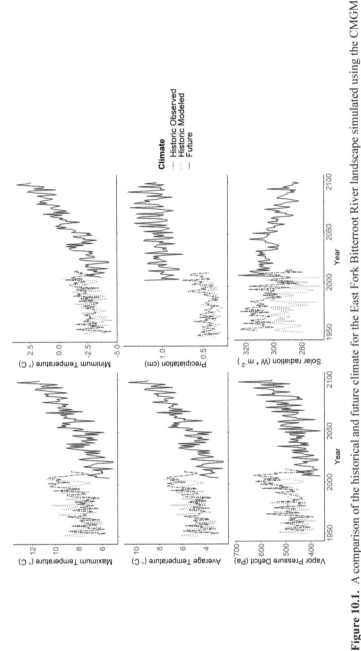

Figure 10.1. A comparison of the historical and future climate for the East Fork Bitterroot River landscape simulated using the CMGM model with the historical observations from the Sula, Montana, USA weather station (Keane et al. 2017). Note three things about this comparison: (1) the historical observed weather (light dotted line) is often in disagreement with the simulated historical weather (dark dotted line), (2) the historical observed weather cycles are not synchronized with historical simulations, and (3) the future predictions (solid line) do not start where the observed or simulated weather end. This makes it difficult to calibrate and execute the model using only simulated climate sources.

that any comparisons of historical to future predictions should use only simulated data because both simulated weather streams contain the same levels of uncertainty. This may be difficult for many users because model calibration used observed weather, and those weather observations represent the major influences of the teleconnections that drive climate cycles, such as El Niño, *La Niña*, and the Pacific Decadal Oscillation (Newman et al. 2003). As a result, simulations with observed weather will more accurately match observed ecosystem responses than the gridded simulated weather reconstructions. In some of my modeling projects, we've found an alternative approach may be to compare the last decade of future predictions to the last decade of actual observations to determine offset factors that are then used to adjust the observed values based on the simulated time span (Keane et al. 1996, Holsinger et al. 2014).

Another challenge for ecological modelers is that weather streams required as inputs to models often contain variables that were never collected; temperature and precipitation, for example, were measured but solar radiation or wind speed and direction might be missing. There are a couple of reasons why these weather data are missing: (1) a variable may have missing vales in the input stream and/or (2) a variable may not appear in the input stream because it wasn't measured. These two situations happen frequently for weather station observations. Methods must be developed to deal with these data limitation problems. Interpolation may be possible with missing weather values (e.g., use the previous 10 days to estimate the missing value), and correlation might be an alternative with missing variables (compute missing value from regression equation using other measured values) (Bristow and Campbell 1984). More often, missing variables can be obtained from other sources, preferably a nearby location where the original data was measured. As an example, most weather observations in the US commonly have only temperature and precipitation, therefore Keane et al. (2011) employed the MTCLIM model (Hungerford et al. 1989) to calculate radiation, humidity, and day-length from weather observations from a National Weather Service weather station to run the FireBGCv2 model. In later projects, we used DAYMET (Thornton et al. 1997) and PRISM (Daly et al. 2008) to obtain fine-scale weather observations in complex terrain (Keane and Holsinger 2006).

A last challenge is the scale mismatch between the input climate data and the model's simulation resolution. Often, climate grid data may be too coarse for use in small landscape or stand applications because fine-scale interactions of climate with topography and vegetation are missing when the climate grid resolutions are coarse (Keane and Holsinger 2006). Adiabatic lapse rates, for example, may not be properly represented in the climate grid when the pixel size is so large that fine scale variations in topography are not included in the climate simulations (Thornton et al. 2000). In mountainous terrain, a coarse climate grid may have the same daily weather calculations for adjacent grids even though there may be a 3000 m difference in elevation. This confounds many fine scale ecological simulations, especially in remote areas with few ground-based weather stations, because the weather inputs may inaccurately represent climate dynamics for a given landscape. Adjusting grid values based on a statistical comparison to local or nearby weather observations may be the only way to obtain realistic weather and climate inputs across highly dissected landscapes (Keane and Holsinger 2006).

Disturbance

As mentioned in previous chapters, one of the biggest limitations of most ecological models is that they rarely simulate the dynamics of all of the disturbances that can occur on a simulation landscape; they only simulate dynamics of a selected set of disturbances that the modeler felt were important enough to include in the model. This is, of course, a necessary simplification of an overly complex and complicated system, and it also is done because information is often lacking for many disturbance agents (e.g., mistletoe, drought, and windthrow). This simplification may be acceptable to both modeler and users under today's climate conditions, but, in the uncertain climate future, it is difficult to know if an un-modeled disturbance may become a major factor in dictating future ecosystem dynamics. Moreover, there may be some management practices that, when implemented, would facilitate the future initiation and spread of a disturbance that was not included in the model. The lack of other biophysical processes in model design is also important, not just disturbance; sublimation, for example, may not be included in a model that simulates snow dynamics but sublimation may be an important water flux in the future. In one project, Keane et al. (2015a) explored interactions across three disturbances in a modeling project, but two of the three simulated disturbances only impacted shade-intolerant species even though the landscape was dominated by shade tolerant species in the future. Impacts of disturbance agents and other biophysical processes that were not simulated must somehow be factored into the interpretation of the results.

Users should also pay special attention as to how their selected model simulates the four stages of a disturbance event (Chapter 2). It is somewhat important that the rigor of disturbance simulation be similar across all four stages and that each stage is represented in the simulation. For example, the simulation of fire spread may be highly complex and mechanistic yet the computation of the effects of fire on the ecosystem may be overly simplistic. The representation of each disturbance stage in the selected model is important because it will factor into how users will evaluate model results. It will be important, for instance, to know that fire-caused tree mortality is computed from a three category classification of fire severity rather than a more mechanistic process that recognizes a tree's species and size, and the intensity of the fire. In keeping with the ignition examples presented in previous sections, it may be important to know if ignition locations are tied to fuels and topography, or are they simply random across the simulated area. Spread dynamics are also important because the detail at which a disturbance agent proceeds across the landscape can dictate future patterns and processes; cell automata simulations of fire spread, for example, may create fire burn patterns that are rarely found on today's landscapes (Andrews 1989, Ball and Guertin 1992, Finney 1998).

Another issue that often crops up during calibration and analysis is when the simulated disturbance regimes fail to match expected regimes. Often, information on a disturbance regime has been collected from an area outside of the simulation landscape, but for a similar set of vegetation types, biophysical settings, or structural stages. These data are used to quantify model parameters and to compare with model results. Unfortunately, each landscape is unique in terms of topography, soils, weather, and vegetation conditions. Therefore, the topographic structures and vegetation compositions of the simulated landscape may be quite different from the landscape where the disturbance regime parameter data were collected. This may result in the characteristics of the simulated landscape overwhelming the influence of the parameters whose values were estimated elsewhere. Wildland fire regimes, for example, are often described using a fire return interval and an average fire size (Agee 1993), and many models incorporate these parameters into algorithms to simulate fire spread (Li 1997, Sturtevant et al. 2009). However, values for these parameters are often estimated from data collected in areas outside of the simulation area and for areas that may have are not representative of the simulation area. For example, fire history studies may collect data from flat areas and these data are then used to parameterize fire regimes for highly dissected terrain, yet we know that fire spreads much differently in flat vs mountainous areas. Users can accept the fact that the simulated fire regime is different from that reported in the literature and hopefully, they can be reassured that the model is simulating the effects of topography and weather on disturbance dynamics.

HRV simulations

There are several issues that are involved when ecological models are used to quantitatively describe the range and variation of historical disturbance and vegetation regimes (HRV) by simulating thousands of years of ecosystem response (**HRV approximations**). First, people often criticize this approach because the model did not use the true thousand year climate record and the model was not parameterized and initialized for the entire thousand years time slice. They are right to assume that the cycling of a limited weather record for 1,000 years does not fully describe historical climate conditions over that millennia. However, the primary purpose of the HRV modeling efforts is to describe *variability* in historical landscape dynamics, not to determine actual landscape conditions for each historical year. To capture this variability, field-based study findings are used to quantify parameters for the models, even though these data may represent a relatively small slice of time (300 to 400 years). The assumption here is that this small temporal data span is a good proxy for the creation of reference conditions used in HRV simulation. Since the field-based time slice represents only three or four centuries, it may seem that only 500 years of simulation are needed. However, the sampled fire events that occurred during this time represent only one unique sequence of disturbance initiation locations, and the subsequent weather that caused the disturbance to spread. If these events had happened on a different timetable or in different areas, an entirely new set of landscape conditions would have resulted. It follows then that the documentation of landscape conditions from only historical records would tend to underestimate the variability of conditions that landscape could have experienced in the past. As a result, the entire range of historical conditions may be approximated by simulating the historical disturbance regime for thousands of years assuming that a thousand years are long enough to describe the complete manifestation of all disturbance events.

A second issue is the appropriateness of simulation for quantifying HRV—many feel that modeling involves such oversimplification that the results may be inappropriate for accurately estimating HRV (Keane et al. 2009). This may be true in some cases, such as landscapes with extensive data and published information. But what are the alternatives for other areas? Chronosequences of historical data are mostly absent for most HRV applications, and even if the data were available, they would be limited in temporal and spatial scale (Humphries and Bourgeron 2001). Spatial chronosequences (i.e., estimating variability by comparing maps or conditions across similar landscapes) (Hessburg et al. 1999) have a great deal of uncertainty as well and don't really capture historical variation (Keane et al. 2009). Simulation modeling may be the only viable alternative for constructing useful and meaningful HRV time series (Keane 2012). There are several advantages of using modeling for HRV quantification. First, the same model can be used to generate other ranges of variation, such as future ranges under climate change and management actions (Keane et al. 2018), that can be integrated with HRV to identify zones of overlap. Next, the model can be used to approximate the entire range of possible range of conditions that could have happened, rather than just the set of conditions that actually happened. The model can also be tuned to generate conditions along any segment of historical timelines, from recent to paleo dynamics (Prentice et al. 1991). And, the model can be used to identify those aspects of historical ecology that need further study. Many projects have augmented historical records with simulated

198

historical conditions to more accurately define HRV. Ecological modeling may provide the "safety net" for most HRV projects.

Modeling science

Many people believe that complex models are somehow "better" than simple, parsimonious models because they appear to have more under the "hood". Conversely, others feel that complicated models have so many moving parts that they are impossible to parameterize correctly and, as a result, often yield erroneous results because errors across interacting modules often multiply and compound (Perry and Millington 2008). After a career of oscillating back and forth between these two camps, I have come to realize that it is nearly impossible to compare existing models for quality without a well-defined context, which is often arbitrary and centered around unique modeling objectives. All models have their place in any given application; managers with limited computer resources, for example, may not be able to execute highly complex models if thousands of runs are needed. Or, a set of specific variables needed by managers for land planning may only be available from a set of complex models. Therefore, the selection and evaluation of models shouldn't always prioritize model design and detail, but rather they should also take into account the original reasons for which the model was built (i.e., the original modeling objective) and how the model will be used (i.e., the current objective). Comparing individual models is a bit like comparing apples and oranges because they are so different in scale, scope, and application (Keane et al. 2004), yet all of them may have their place in the management decision space.

There are ample comparisons of different ecological models in the literature, and these evaluations provide a wonderful means of understanding the relative strengths and weaknesses of each model (Amthor et al. 2001, Barrett 2001). However, these comparisons should be carefully evaluated to determine if they are really relevant for the project's modeling objective. Some comparisons may apply a model to situation for which it wasn't designed, or the initialization and parameterization may not be of the same rigor across evaluated models. While model comparison studies are informative, they often have great inconsistencies that may not provide adequate evidence that a user's selected model is either good or bad.

199

Therefore, user's should evaluate comparison results in the context of the comparison only; for example, were the same data used for all models, or were the models properly calibrated, or was there sufficient time spend to properly run each model?

Another issue that is often important to some modeling projects is whether the results from the model were emergent (resulting from complex interactions of diverse simulated processes) or if the results were a direct consequence of the initialization and parameterization of the model (Wilson and Botkin 1990, Olson and Sequeira 1995, Stewart et al. 2013). In other words, can model results be estimated from the parameterization, or are the results dictated by intricate interactions across numerous modules. This is important because emergent results often lead to novel ecosystem behaviors, unexpected outcomes, and interesting insights into ecological systems, whereas models with largely deterministic results are simply large, complex calculators. An example would be the simulation of a landscape's fire regime using an S&T model vs a spatially explicit mechanistic ecosystem model with wildland fire modules. In the S&T model, fire occurrence probabilities are often input parameters, and therefore, fire return intervals for the simulated landscape can be easily estimated prior to simulation from probability values. However, in more complex spatial models, fire return intervals might result from stochastic ignition processes coupled with mechanistic fire growth simulations, complex fire effects calculations, and diverse vegetation development pathways, all of which are dictated by topography, heterogeneous fuel mosaics, and climate. As a result, the fire return interval is an emergent property resulting from multifaceted interactions. In land management, it may not be that important to integrate climate, vegetation, topography and disturbance to obtain a prediction, such as simulating timber volume over the next 20 years, but some projects may need to explore changes in fire regime as a result of warming climates and changing land management activities, and therefore it is important that the simulated fire regimes are emergent due to the complexities of model design.

Some users don't fully understand why it is so difficult to predict landscape conditions into the future. A thought experiment may serve to demonstrate why this is so hard, and it might provide a broader context in which to understand and interpret ecological models. Pretend a modeling project is seeking to predict landscape or ecosystem conditions some 50 years from now. It is somewhat straightforward to simulate vegetation dynamics for that short time span, and many landscape models can do this quite well. However, the biggest obstacle preventing accurate predictions is the simulation of major exogenous (outside the simulation area) or endogenous (inside the area) disturbances. How does a stand or point model anticipate the timing and effects of an exogenous or endogenous disturbance? And, if a disturbance is spatially simulated in an LESM, the determination of the exact location, spread, intensity, and severity of disturbance is extremely difficult and subject to many fine-scale factors that are difficult to realistically represent and replicate in a model. Wildland fire, for example, often starts from a point (e.g., lightning strike) and spreads outwards along vectors of slope and wind. Prediction of the exact location of that lightning strike is nearly impossible because of the high stochasticity of lightning dynamics (Barrows et al. 1977), and subsequent spread is difficult because of the timing of the ignition determines fuel moistures and more importantly, highly variable wind fields. As a result, the future prediction of landscape or ecosystem conditions would rely heavily on where that fire started and where the fire spread making it is difficult to forecast conditions 50 years in the future. Because of this great uncertainty, many feel the most appropriate use of modeling is to simulate regimes rather than a specific disturbance event to approximate a range of future conditions that capture the variability for all possible fire starts.

Ensemble modeling has been used to reduce uncertainty and increase confidence in some modeling projects (Thuiller et al. 2009, Keane et al. 2013a). These comparative approaches use several models to simulate the same set of scenarios and the subsequent results are compared across models to determine if they are comparable across all models (Horn 1966, Mellert et al. 2015). The assumption is that if a simulated result is similar across one or more models, it is more likely to be realistic or accurate. I was involved in a 25 year project that used a set of 3–5 models to simulate climate change and land management impacts on landscape fire dynamics and we learned a great deal about fire and vegetation dynamics using the ensemble approach (Cary et al. 2006, Keane et al. 2007, Cary et al. 2009, Keane et al. 2013a, Cary et al. 2016). One way that ensemble modeling can help in operational applications is to evaluate the uncertainty in a selected model (e.g., if the selected model was used in an ensemble project and its results were similar to other models, then the modeling team can assume that results generated from that model are more generally applicable). Or, if there are two or more models that are available and are easily used to simulate a project's scenarios, then it might be of great benefit to the project to do an ensemble approach. In the ensemble projects that I've been involved, we found a consensus among models with respect the relative importance of some important factors (e.g., climate, wildfire management) over others (e.g., fuel treatment effort). Ensemble results can genuinely influence operational decisions by identifying the most important factors that determine impacts of various alternatives. However, while providing great insights into the relative importance of factors governing fire dynamics in landscapes, the time and expense of initializing, parameterizing, and calibrating more than one model under an unbiased simulation design may be impractical for some natural resource projects.

Summary

The purpose of this book was to help neophyte ecological modelers implement a complex modeling project with a high degree of confidence and interpret model results and uncertainties in an appropriate manner. Some of the material in this book may not be completely applicable to a specific modeling project because the details of model preparation are often dictated by local conditions, or some of the modeling phases may have already been done by others. Admittedly, this book emphasizes mechanistic landscape ecological models and their application in natural resource management so some of the steps presented may be immaterial when other ecological models are employed. Moreover, the steps and level of detail presented in each chapter may not be relevant to the project objective, selected model, or desired output resolution. Hopefully, there is sufficient material in this book to conduct a successful modeling campaign to achieve desired simulation objectives. Good modeling.

References

Agee, J. K. 1993. Fire ecology of Pacific Northwest forests. Island Press, Washington DC USA.

Amthor, J. S., J. M. Chen, J. S. Clein, S. E. Frolking, M. L. Goulden, R. F. Grant, J. S. Kimball, A. W. King, A. D. McGuire, N. T. Nikolov, C. S. Potter, S. Wang and S. C. Wofsy. 2001. Boreal forest CO2 exchange and evapotranspiration predicted by nine ecosystem process models: Intermodel comparisons and relationships to field measurements. Journal of Geophysical Research: Atmospheres 106: 33623–33648.

Andrews, P. L. 1989. Application of fire growth simulation models in fire management. pp. 317–321. *In*: 10th Conference on Fire and Forest Meteorology, Ottawa, Canada.

Ball, G. L. and D. P. Guertin. 1992. Advances in fire spread simulation. pp. 241–249. *In*: Proceedings on the Third Forest Service Remote Sensing Applications Conference— Protecting natural resources with remote sensing. American Society of Photogrammery and Remote Sensing, 5410 Grosvenor lane,Bethesda, Maryland USA, Tucson, Arizona USA.

Barrett, T. M. 2001. Models of vegetative change for landscape planning: A comparison of FETM, LANDSUM, SIMPPLLE, and VDDT. General Technical Report RMRS-GTR-76-WWW, USDA Forest Service, Rocky Mountain Research Station, Ogden, UT, USA.

Barrows, J. S., David V. Sandberg and J. D. Hart. 1977. Lightning fires in Northern Rocky Mountain forests. Final Report for Contract Grant 16-440-CA, USDA Forest Service Intermountain Fire Sciences Laboratory, On file, USDA Forest Service, Intermountain Fire Sciences Laboratory, P.O. Box 8089, Missoula, MT., USA.

Box, G. E. P. and G. C. Tiao. 1975. Comparison of forecast and actuality. Applied Statistics 25: 195–200.

Bristow, K. L. and G. S. Campbell. 1984. On the relationship between incoming solar radiation and daily maximum and minimum temperature. Agricultural and Forest Meteorology 31: 159–166.

Cary, G., M. D. Flannigan, R. E. Keane, R. Bradstock, I. D. Davies, J. L. Lenihan, C. Li, K. Logan and R. Parsons. 2009. Relative importance of fuel management, ignition likelihood, and weather to area burned: evidence from five landscape fire succession models. International Journal of Wildland Fire 18: 147–156.

Cary, G. J., I. D. Davies, R. A. Bradstock, R. E. Keane and M. D. Flannigan. 2016. Importance of fuel treatment for limiting moderate-to-high intensity fire: Findings from comparative fire modelling. Landscape Ecology: 1–11.

Cary, G. J., R. E. Keane, R. H. Gardner, S. Lavorel, M. D. Flannigan, I. D. Davies, C. Li, J. M. Lenihan, T. S. Rupp and F. Mouillot. 2006. Comparison of the sensitivity of landscape-fire-succession models to variation in terrain, fuel pattern, climate and weather. Landscape Ecology 21: 121–137.

Daly, C., M. Halbleib, J. I. Smith, W. P. Gibson, M. K. Doggett, G. H. Taylor, J. Curtis and P. P. Pasteris. 2008. Physiographically sensitive mapping of climatological temperature and precipitation across the conterminous United States. International Journal of Climatology 28: 2031–2064.

Finney, M. A. 1998. FARSITE: Fire Area Simulator—model development and evaluation. Research Paper RMRS-RP-4, United States Department of Agriculture, Forest Service Rocky Mountain Research Station, Ft. Collins, CO USA.

203

Gustafson, E. 2013. When relationships estimated in the past cannot be used to predict the future: using mechanistic models to predict landscape ecological dynamics in a changing world. Landscape Ecology 28: 1429–1437.

Gustafson, E. J., A. Z. Shvidenko, B. R. Sturtevant and R. M. Scheller. 2010. Using landscape disturbance and succession models to support forest management. *In*: C. Li, R. Lafortezza and J. Chen (eds.). Landscape Ecology and Forest Management: Challenges and Solutions in a Change Globe. Springer International Publishing, Switzerland.

Hessburg, P. F., B.G. Smith and R. B. Salter. 1999. A method for detecting ecologically significant change in forest spatial patterns. Ecological Applications 9: 1252–1272.

Holsinger, L., R. E. Keane, D. J. Isaak, L. Eby and M. K. Young. 2014. Relative effects of climate change and wildfires on stream temperatures: A simulation modeling approach in a Rocky Mountain watershed. Climatic Change 124: 191–206.

Horn, H. S. 1966. Measurement of overlap in comparitive ecological studies. The American Naturalist 100: 419–429.

Humphries, H. C. and P. S. Bourgeron. 2001. Methods for determining historical range of variability. pp. 273–291. *In*: M. E. Jensen and P. S. Bourgeron (eds.). A Guidebook for Integrated Ecological Assessments. Springer-Verlag, New York, New York, USA.

Hungerford, R. D., R. R. Nemani, S. W. Running and J. C. Coughlan. 1989. MTCLIM: A mountain microclimate simulation model. Research Paper INT-414, USDA Forest Service, Intermountain Research Station, Ogden, UT.

Keane, R. E. 2012. Creating historical range of variation (HRV) time series using landscape modeling: overview and issues. pp. 113–128. *In*: J. A. Wiens, G. D. Hayward, H. S. Stafford and C. Giffen (eds.). Historical Environmental Variation in Conservation and Natural Resource Management. John Wiley and Sons, Hoboken, New Jersey.

Keane, R. E., G. Cary, I. D. Davies, M. Flannigan, R. H. Gardner, S. Lavorel, J. M. Lenihan, C. Li and T. S. Rupp. 2007. Understanding global fire dynamics by classifying and comparing spatial models of vegetation and fire dynamics. *In*: J. Canell, L. Patalki and P. P. (eds.). Terrestrial Ecosystems in a Changing World—GCTE Synthesis Book. Cambridge University Press, Cambridge, United Kindom.

Keane, R. E., G. Cary, I. D. Davies, M. D. Flannigan, R. H. Gardner, S. Lavorel, J. M. Lennihan, C. Li and T. S. Rupp. 2004. A classification of landscape fire succession models: Spatially explicit models of fire and vegetation dynamic. Ecological Modelling 256: 3–27.

Keane, R. E., G. J. Cary, M. D. Flannigan, R. A. Parsons, I. D. Davies, K. J. King, C. Li, R. A. Bradstock and M. Gill. 2013a. Exploring the role of fire, succession, climate, and weather on landscape dynamics using comparative modeling. Ecological Modelling 266: 172–186.

Keane, R. E., J. M. Herynk, C. Toney, S. P. Urbanski, D. C. Lutes and R. D. Ottmar. 2013b. Evaluating the performance and mapping of three fuel classification systems using Forest Inventory and Analysis surface fuel measurements. Forest Ecology and Management 305: 248–263.

Keane, R. E., P. F. Hessburg, P. B. Landres and F. J. Swanson. 2009. A review of the use of historical range and variation (HRV) in landscape management. Forest Ecology and Management 258: 1025–1037.

Keane, R. E. and L. Holsinger. 2006. Simulating biophysical environment for gradient modeling and ecosystem mapping using the WXFIRE program: Model documentation and application. Research Paper RMRS-GTR-168CD, USDA Forest Service Rocky Mountain Research Station, Fort Collins, Co, USA.

Keane, R. E., L. Holsinger, M. F. Mahalovich and D. F. Tomback. 2017. Restoring whitebark pine ecosystems in the face of climate change. General Technical Report RMRS-GTR-361. USDA Forest Service Rocky Mountain Research Station, Fort Collins, CO. 123 pages.

Keane, R. E., R. Loehman, J. Clark, E. Smithwick and C. Miller. 2015a. Exploring interactions among multiple disturbance agents in forest landscapes: Simulating effects of fire, beetles, and disease under climate change. pp. 201–231. *In*: A. H. Perera, T. K. Remmel and L. J. Buse (eds.). Modeling and Mapping Forest Landscape Patterns. Springer, New York, USA.

Keane, R. E., R. A. Loehman and L. M. Holsinger. 2011. The FireBGCv2 landscape fire and succession model: A research simulation platform for exploring fire and vegetation dynamics. General Technical Report RMRS-GTR-255, U.S. Department of Agriculture, Forest Service, Rocky Mountain Research Station, Fort Collins, CO USA.

Keane, R. E., R. A. Loehman, L. M. Holsinger, D. A. Falk, P. Higuera, S. M. Hood and P. F. Hessburg. 2018. Use of landscape simulation modeling to quantify resilience for ecological applications. Ecosphere 9: e02414.

Keane, R. E., D. McKenzie, D. A. Falk, E. A. H. Smithwick, C. Miller and L.-K. B. Kellogg. 2015b. Representing climate, disturbance, and vegetation interactions in landscape models. Ecological Modelling 309-310: 33–47.

Keane, R. E., K. C. Ryan and S. W. Running. 1996. Simulating effects of fire on northern Rocky Mountain landscapes with the ecological process model Fire-BGC. Tree Physiology 16: 319–331.

Li, C. 1997. ON-FIRE: A landscape model for simulating the fire regime of northwest Ontario. Ecological Research and Sustainable Development 4: 369–392.

Mellert, K. H., V. Deffner, H. Küchenhoff and C. Kölling. 2015. Modeling sensitivity to climate change and estimating the uncertainty of its impact: A probabilistic concept for risk assessment in forestry. Ecological Modelling 316: 211–216.

Menakis, J. P., Robert E. Keane and D. G. Long. 2000. Mapping ecological attributes using an integrated vegetation classification system approach. Journal of Sustainable Forestry 11: 245–265.

Millar, C. I., N. L. Stephenson and S. L. Stephens. 2007. Climate change and forests of the future: Managing in the face of uncertainty. Ecological Applications 17: 2145–2151.

Newman, M., G. P. Compo and M. A. Alexander. 2003. ENSO-Forced Variability of the Pacific Decadal Oscillation. Journal of Climate 16: 3853–3857.

Olson, R. L. and R. A. Sequeira. 1995. An emergent computational approach to the study of ecosystem dynamics. Ecological Modelling 79: 95–120.

Perry, G. L. W. and J. D. A. Millington. 2008. Spatial modelling of succession-disturbance dynamics in forest ecosystems: Concepts and examples. Perspectives in Plant Ecology, Evolution and Systematics 9: 191–210.

Prentice, I. C., P. J. Bartlein and T. Webb III. 1991. Vegetation and climate change in eastern North America since the last glacial maximum. Ecology 72: 2038–2056.

Rupp, D. E., J. T. Abatzoglou, K. C. Hegewisch and P. W. Mote. 2013. Evaluation of CMIP5 20th century climate simulations for the Pacific Northwest USA. Journal of Geophysical Research: Atmospheres 118: 10,884–810,906.

Schmolke, A., P. Thorbek, D. L. DeAngelis and V. Grimm. 2010. Ecological models supporting environmental decision making: a strategy for the future. Trends in Ecology & Evolution 25: 479–486.

Stewart, J., A. J. Parsons, J. Wainwright, G. S. Okin, B. T. Bestelmeyer, E. L. Fredrickson and W. H. Schlesinger. 2013. Modeling emergent patterns of dynamic desert ecosystems. Ecological Monographs 84: 373–410.

Sturtevant, B. R., R. M. Scheller, B. R. Miranda, D. J. Shinneman and A. Syphard. 2009. Simulating dynamic and mixed-severity fire regimes: A process-based fire extension for LANDIS-II. Ecol. Model 220.

Taylor, K. E., R. J. Stouffer and G. A. Meehl. 2012. An Overview of CMIP5 and the Experiment Design. Bulletin of the American Meteorological Society 93.

Thornton, P. E., H. Hasenauer and M. A. White. 2000. Simultaneous estimation of daily solar radiation and humidity from observed temperature and precipitation: An application over complex terrain in Austria. Agricultural and Forest Meteorology 104: 255–271.

Thornton, P. E., S. W. Running and M. A. White. 1997. Generating surfaces of daily meteorological variables over large regions of complex terrain. Journal of Hydrology 190: 214–251.

Thuiller, W., B. Lafourcade, R. Engler and M. B. Araújo. 2009. BIOMOD–a platform for ensemble forecasting of species distributions. Ecography 32: 369–373.

Wilson, M. V. and D. B. Botkin. 1990. Models of simple microcosms: Emergent properties and the effect of complexity on stability. The American Naturalist 135: 414–434.

Index

A

Accuracy 8, 16, 17, 23, 24, 28, 58, 61, 62, 67, 72, 90, 114, 133–135, 140, 145, 147–149, 151–153, 191
Algorithm 6, 12, 13, 15, 18–22, 28, 30–32, 72, 94–97, 100, 101, 103, 107, 109, 115, 119, 126, 128, 131, 137, 140, 153, 158, 197
ANOVA 75, 167, 171, 183, 184
Appropriateness 63, 198
Association 137, 174, 191

B

Bias 102, 145–147, 167
Biophysical setting 86, 196
Box and whisker diagram 167

C

Calibration 8, 18, 22–24, 59, 60, 65, 67, 77, 82, 87, 91, 96–98, 101, 103, 108, 113–119, 121–131, 133, 134, 141, 142, 158, 159, 162–164, 168, 194, 196
Calibration strategy 113, 116, 118, 121, 122, 126, 129
Climate 2–4, 6, 9, 12, 22, 25, 26, 31, 36–44, 57, 58, 60, 61, 73–76, 82, 94, 116, 117, 122, 123, 130, 136, 167, 170–173, 175

Climate change 2, 25, 26, 39, 43, 44, 57, 184, 189, 198, 201
Cloud computing 162
Community 1, 20, 35, 36, 44, 66, 83, 110, 145
Comparative modeling study (*see* Ensemble modeling)
Computer memory 59, 164
Contingency table 138, 147, 148
Cover type 20, 35, 36, 66, 86, 91, 139, 167, 191

D

Daily 29, 35, 40–42, 61, 74, 152, 192, 195
Data 3, 5, 6, 8, 13, 14, 16, 19, 22–26, 30, 40–42, 44, 53, 61–63, 66, 68, 69, 80–87, 90, 95–99, 101, 114, 116, 119, 121, 122, 125, 129, 130, 133–137, 139–148, 151, 152, 160, 162, 163, 165–171, 174, 175, 179, 180, 183, 184, 186, 189, 194–198, 200
Data existing 102, 136, 142, 143
Data field 89
Data field campaign 87
Data format 88, 97, 134, 144
Data layers 83, 86, 87
Data legacy 95
Data plot 86, 87
Data remotely sensed 99, 143
Decadal 35, 41, 194
Design factor 75–76

Design level 75–76
Disk storage 161, 182
Disturbance 3, 6, 9, 26, 31, 34, 38, 39, 43, 58, 60, 69, 70, 72, 82, 103, 104, 115–118, 122, 125, 127, 167, 188, 192, 195, 196–198, 200, 201
Disturbance effects 38
Disturbance initiation 43, 72, 198
Disturbance spread 34, 72, 192
Disturbance termination 38
Documentation 85, 104, 108, 164, 198
Dynamic Global Vegetation Models (DGVM) 5, 34, 36

E

EFBR 45, 74, 172, 173, 175–180, 185, 192
Emergent results 200
Empirical analysis 189
Ensemble modeling 201
Execution 19, 22–24, 55, 63–65, 75, 77, 89, 113–115, 129, 157–164, 170
Expert opinion 95, 96, 101, 103, 189
Expertise 1, 2, 5, 9, 13, 26, 53, 55, 59, 60, 63, 65, 68, 76, 77, 99, 105, 118, 125, 129, 189

F

Factorial design 22, 54, 75, 150, 151, 167
Field crew 25
Figure 3, 4, 6–8, 16–23, 25, 27–29, 33, 35, 37, 43, 45, 54–58, 64, 70–72, 74, 84, 85, 90–92, 106, 116, 117, 119–124, 129–131, 141, 146–148, 153, 158–160, 162, 165, 166, 168–171, 174–182, 184, 185, 192, 193
File directory 159–160
FireBGCv2 4, 21, 35, 37, 39, 41, 45, 83, 86, 139, 140, 160, 176, 177, 178, 180, 185, 194
Forest Inventory and Analysis (FIA) 86, 87, 98, 144

FRAGSTATS 180
Function 18, 19, 28, 30, 35, 36, 103, 105, 119, 122, 170

G

GIS 59, 86, 179, 180
Grain 20

H

Half-double 128
Historical Range and Variation (HRV) 44, 174–176, 197–199
HRV approximations 197

I

Imagery 99, 143
Initial value 16, 21, 80–82, 84, 89, 91, 101
Initialization 7, 14, 16, 18, 19, 22, 23, 54, 55, 58, 59, 61, 63, 65, 77, 80–92, 97, 98, 103, 110, 113, 115, 116, 125, 134, 141, 142, 158, 159, 162–164, 189, 199, 200
Initialization scheme 86
Intercept 147
Issues 6, 8, 13, 60, 69, 114, 130, 145, 165, 170, 188,
Iteration 104, 105, 128
Iterative process 127, 128

L

LANDFIRE 87
Landsat 99, 143
Landscape 84, 86
Landscape buffer 70, 192
Landscape composition 20, 31, 32, 86
Landscape Ecosystem Simulation Models (LESMs) 6, 12, 13, 22, 30, 32, 35, 45, 67,69, 86, 90, 96, 103, 104, 114, 135
Landscape structure 20, 180
Layer 18, 83, 86, 87, 139

Least square regression 30
Lidar 99

M

Management actions 1–4, 6, 13, 73, 75,
    87, 118, 184, 189, 191, 198
Management alternatives 22, 43, 54, 57,
    73
Map 20, 22, 24, 86, 179, 180, 181
Mapscale 20, 26
Metadata 89
Model agent based (ABM) 4, 42
Model approximation 95–97, 100–103,
    107
Model behavior 8, 24, 72, 113–118, 121,
    122, 126, 130, 131, 141, 149, 151,
    159, 179
Model climate 4, 6, 12, 39–42, 94, 192
Model cohort 35, 37
Model complex 14, 25, 27, 28, 32, 58–60,
    64–66, 96, 98, 102, 103, 114, 153,
    160, 199, 202
Model deterministic 28
Model ecosystem 1, 5, 12, 14, 15, 23, 31,
    33, 38, 39, 60, 64, 65, 73, 85, 135,
    164, 200
Model empirical 29, 30
Model equilibrium 30, 82, 83
Model exploratory 31
Model fire 31, 34, 67, 97
Model gap 37, 96, 105, 116
Model Global Circulation (GCM) 40, 41,
    73, 192
Model growth and yield 5, 30, 139, 144
Model individual based (IBM) 4, 42
Model landscape 13, 31, 34, 35, 40, 45,
    53, 60, 61, 72, 81, 83, 84, 87, 115,
    116, 152, 179, 180, 192, 201
Model management 32, 86
Model mechanistic 30, 42, 58, 82, 96,
    131, 137
Model non-equilibrium 30

Model non-spatial 30, 31
Model plant 35, 37
Model point 34, 69, 201
Model population 42
Model prognostic 31
Model regional 40
Model research 32, 76, 89
Model selection 62–68
Model simple 25, 58, 59, 64, 104, 139
Model spatial 30, 31, 34, 63, 135, 151,
    200
Model spatially explicit 6, 30, 192
Model species 35
Model state and transition (S&T) 35, 36,
    60, 139, 147, 167, 191, 200
Model statistical 5, 30, 41, 101
Model stochastic 28, 29, 31, 32, 67, 72,
    145, 152, 153
Modeler 13, 15–17, 24, 25, 57, 73, 84, 85,
    89, 110, 151, 158, 166, 190, 191, 195
MODIS 99, 139, 140, 143
Module 18, 19, 39, 64, 72, 95, 108, 137,
    140, 143, 144, 199, 200
Monthly 40, 61, 66, 140
MTCLIM 41, 194

N

natural range and variation 16, 17, 153
Natural resource management 2, 3, 5, 6,
    8, 9, 12, 24, 34, 36, 43, 55, 56, 63,
    83, 202
neutral Landscape 83
NPP 13, 139, 140, 143

O

Objectives 8, 14, 15, 21, 53, 55, 56, 60,
    61, 63, 64, 66, 69, 71–73, 78, 81, 107,
    118, 134, 166, 167, 171, 182, 190,
    199, 202
Observed values 135, 137, 145, 146, 194
Operating systems 67, 163

P

Parameter 7, 14, 18, 21–23, 25, 32, 55, 94, 96–110, 114–115, 118, 119, 121, 122, 127–131, 134, 142, 149, 150, 151, 164, 197

Parameterization 7, 14, 16–18, 22, 23, 54, 55, 58, 61–63, 65, 77, 87, 90, 94–96, 98–110, 113, 115, 116, 122, 125, 134, 141, 142, 151, 153, 158, 159, 162, 163, 164, 170, 189, 199, 200

Parameterization method 96

Parameterization strategy 95, 102,103, 105, 106, 108

Patch 20, 34, 152, 184

Pixel 20, 30, 31, 40, 42, 61, 86, 95, 140, 195

Plant functional type (PFT) 35, 36, 103–105

Point 15–17, 22, 24, 27, 31, 34, 41, 69, 76, 77, 90, 95–97, 104, 108, 109, 114, 130, 135, 143, 145, 148, 158, 166, 175, 176, 191, 201

Precision 14, 16, 17, 23, 114, 134, 140, 147, 149

Predicted values 135, 145–147

Principal Components Analysis (PCA) 174, 176

PRISM 194

Processor speed 63, 163

Productivity gross 152

Productivity net primary (NPP) 136, 139, 140, 143

Programmer 25

Project design 6, 15, 22, 53, 54, 57, 66, 72, 75, 96, 158, 161, 164, 188, 190

Project journal 130

Project time lines 64, 90, 110

Prototype figures 72

Prototype tables 72

R

Raster 20, 27, 31, 32, 40, 86

Receiver operator curve (ROC) 148, 149

Region 4, 41, 81, 180

Regression analysis 146, 147, 151

Replicates 29, 54, 72, 76, 157, 158, 160–164, 171, 179, 183, 184

Residual analysis 145

Resolution 6, 16, 20, 28, 31, 35, 40–42, 51, 53, 60–63, 66–68, 85, 86, 90, 135, 136, 191, 195, 202

Response space 23, 151

Root mean square error (RMSE) 146, 147

S

Sampling field 88, 97, 98, 107, 142

Scale 1, 4, 5, 16, 18, 20, 22, 33–36, 40–42, 60, 61, 85–87, 92, 96, 97, 136, 139, 141, 143, 144, 152, 192, 194, 195, 198, 199, 201

Scatterplot 151, 176

Scenario 6, 14, 22, 41, 43, 54, 55, 57, 58, 60, 61, 64, 65, 67, 73–75, 77, 84, 116–118, 120–123, 126, 127, 129, 130, 157, 158, 160–167, 171–176, 179, 182, 184, 201, 202

Scripts 67, 88, 157, 162, 163

Sensitivity analysis 23, 24, 107, 134, 137, 140, 144, 149–151, 179

simulation Landscape 22, 53, 66, 69–71, 78, 81, 86, 90, 103, 105, 122, 125, 139, 142, 145, 170, 195

Slope 38, 42, 45, 86, 87, 147, 201

SMART principles 54, 55

Software 8, 19, 25, 40, 59, 60 67, 77, 88, 100, 125, 130, 144, 162–165, 168–170, 179, 180, 186

Software Geographic Information Systems (GIS) 59, 86, 179, 180, 186

Spreadsheet 17, 24, 68, 88, 100, 104, 105, 108, 119, 121, 126, 128–130, 164

Stand 4–6, 19, 20–22, 26, 30, 31, 34, 37, 39, 45, 67–69, 74, 82, 84, 86, 137, 144, 145, 152, 180, 191, 195, 201

Statistical package 19, 59, 142, 146, 168, 170, 179, 180

Stochasticity 31, 43, 72, 145, 152, 201
Suitability 63, 64, 191
Supercomputing 161
System 1, 11, 19, 24, 58, 65, 68, 88, 94,
    129, 142, 163, 164, 183, 186, 195

T

Table 18, 25, 26, 27, 58, 62, 63, 73–76,
    96, 97, 104, 105, 119, 121, 122, 126,
    129, 130, 137–139, 141, 148, 153,
    171–174, 176
target Landscape 70, 71, 192
Team 6, 13, 18, 24, 25, 54, 60, 62, 65, 68,
    77, 99, 102, 104, 116–119, 121, 122,
    123, 125–127, 129–131, 133, 134,
    141, 153, 158, 162, 168, 174, 179,
    182, 183, 186, 202
Theory 18, 19, 57, 135, 149
Time series 40, 41, 82, 91, 118, 120, 122,
    123, 125–127, 130, 151, 175–178,
    198
Time span 34, 57, 90, 107, 123, 136, 143,
    144, 184, 194, 201
Time step 34, 35, 74
Timeliness 63, 64
Treatment 67, 73–76, 79, 83, 90, 173, 202
Trial-and-error 82, 114
Tune 94, 113, 198

U

Uncertainty 14–18, 23, 24, 26, 28, 30–32,
    36, 54, 57, 61, 62, 66, 69, 73, 96, 97,
    109, 128, 133–135, 137, 139, 141,
    143, 145, 147, 149, 151, 153, 167,
    189, 191, 192, 194, 198, 201, 202
User 5, 14, 16, 17, 19, 21, 24, 25, 34, 53,
    56, 60, 64, 65, 67, 68, 81, 83, 89,
    90, 92, 94, 95, 99–105, 170–110,
    113, 114, 115, 117–119, 121, 122,
    125–130, 133, 134, 137, 139, 140,
    143, 149–153, 158–164, 166–168,
    170, 179, 182, 184, 190, 191

V

Validation 8, 22, 23, 24, 82, 83, 87, 88,
    96–98, 103, 108, 114, 125, 131,
    133–137, 139, 141–149, 151
Validation approach 142
Variable diagnostic 21, 103, 116, 118, 119,
    121–124, 126, 127, 129, 130
Variable explanatory 21, 57, 118, 139,
    140, 165–167, 175–179
Variable flux 20–22
Variable intermediate 20, 140
Variable output 21, 25, 26, 28, 64, 66, 67,
    107, 116, 145, 164
Variable response 21, 23, 24, 26, 57, 63,
    66, 74, 75, 116, 118, 119, 121–123,
    125, 126, 128, 133, 134, 137,
    139–145, 149, 150, 151, 166, 167,
    171, 172, 174–180, 184
Variable state 20, 21, 24, 35, 42, 63,
    80–82, 84, 85, 192
Variables 8, 20, 23–26, 28, 30, 35, 40,
    41, 44, 56, 57, 60, 63, 64, 66, 67, 74,
    75, 80, 81, 82, 84–86, 100, 103, 107,
    109, 114, 116, 118, 119, 121–127,
    129–131, 133–137, 139–142, 144,
    145, 149–151, 162, 164, 165–167,
    171, 173, 174–180, 182, 184, 191,
    192, 194, 199
Variables input 21, 80, 84, 86
Variables output 21, 25, 26, 28, 64, 67,
    107, 116, 145, 164
Vector 20, 26, 31, 38, 86, 201
Vegetation 5, 6, 9, 18, 20, 31, 34, 35, 36,
    38, 42, 44, 58, 60, 61, 64, 66, 70, 76,
    83, 86, 87, 94, 117, 120, 123, 125,
    152, 170, 174, 195–197, 200, 201

Milton Keynes UK
Ingram Content Group UK Ltd.
UKHW040059071024
449327UK00019B/678

9 780367 779290